CAMBRIDGE STUDIES IN MODERN BIOLOGY: 3

Editors R. S. K. Barnes *Department of Zoology, University of Cambridge*
T. R. Halliday *Department of Biology, The Open University*
P. L. Miller *Department of Zoology, University of Oxford*
D. B. Roberts *Genetics Laboratory, University of Oxford*

DIVING AND MARINE BIOLOGY

To Monica

GEORGE F. WARNER

Department of Zoology, University of Reading

DIVING AND MARINE BIOLOGY

The ecology of the sublittoral

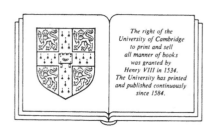

The right of the
University of Cambridge
to print and sell
all manner of books
was granted by
Henry VIII in 1534.
The University has printed
and published continuously
since 1584.

CAMBRIDGE UNIVERSITY PRESS

Cambridge

London New York New Rochelle

Melbourne Sydney

CAMBRIDGE UNIVERSITY PRESS
Cambridge, New York, Melbourne, Madrid, Cape Town, Singapore, São Paulo, Delhi

Cambridge University Press
The Edinburgh Building, Cambridge CB2 8RU, UK

Published in the United States of America by Cambridge University Press, New York

www.cambridge.org
Information on this title: www.cambridge.org/9780521276795

First published 1984
Re-issued in this digitally printed version 2009

A catalogue record for this publication is available from the British Library

Library of Congress Catalogue Card Number: 83–7380

ISBN 978-0-521-25751-0 hardback
ISBN 978-0-521-27679-5 paperback

CONTENTS

PREFACE

Why write a book on diving and marine biology? The answer is obvious until you think about it: all those wonderful experiences gained by confronting sea creatures in their alien environment. But that is the subject of an adventure book; marine biology is a science. Diving, to a marine biologist, is just another method of study; it may be an exciting and aesthetically rewarding method but, as far as the march of scientific progress goes, it is only a method. Thus when one writes scientific papers about research involving diving one does not enlarge upon the mystical experience of penetrating the womb of life, nor even upon the less mystical experiences of getting freezing cold, cut, exhausted, stung or seasick. On the contrary, one presents the findings as concisely as possible and relates them to those of other scientists *who may not have used diving in their studies.* And this is where the difficulty lies: diving is not a marine biological subject, and to achieve an intellectually satisfying treatment one cannot restrict oneself to diving studies. I tried to do so but it wasn't possible. Diving is not even the most up to date research method, providing the answers to all puzzles. It answers some puzzles of course, but mostly it turns up new ones which often require laboratory work for their solution.

There is no doubt, however, that marine biologists who dive bring a special perspective to their work – a feel for the subject that is different from that of non-diving researchers. It is this perspective that I have tried to communicate here by describing from the point of view of a diver the ecology of the environment with which divers are most familiar – the sublittoral, from low tide down to about 60 m. The story of the ecology of the sublittoral is intended to give the book its intellectual cohesion while I hope that the emphasis on diving studies gives it a flavour or feel that is different. Thus the book has two aims and is intended for two types of reader:

divers and non-divers. I hope that both may find useful the gathering into a coherent whole of the scattered information on sublittoral ecology (although I make no claim to have gathered everything). In addition I hope that non-divers máy realize the importance of diving as a research method and may gain some idea of the diver's perspective; the extra insight you get from being there.

In the context of this series of studies in modern biology, diving is a fairly modern research method; not many diving research papers were published before 1970. The sublittoral, especially the rocky sublittoral, is best studied by diving and is therefore a recently investigated habitat. Because of this, modern concepts have been applied to the study of the sublittoral and modern hypotheses tested. I can only hope that I have done justice to the potential of this fruitful subject matter.

ACKNOWLEDGEMENTS

This book is, in many respects, a synthesis of ideas gleaned either from or through others: diving colleagues or diving buddies, conversations at scientific gatherings, correspondence and exchange of publications. I owe a considerable debt of gratitude to all those friends who freely gave their time, their knowledge and their ideas to help me to develop mine. The book is also, however, an account of personal observations, and here the debt of gratitude is to my diving buddies who had to put up with me underwater: fiddling with some experiment, or with my camera, or getting lost in the kelp. Thank you all; I hope that I have made some progress towards justifying your patience and generosity.

I should also like to thank the following who have supplied me with photographic materials: A. R. Ainslie (Fig. 7.3*a*), R. J. A. Atkinson (Fig. 10.10), D. W. J. Bosence (Fig. 10.7*b*), B. E. Brown and J. C. Ogden (Fig. 6.1), C. J. Chapman and J. Main (Fig. 10.9), P. Dustan (Fig. 8.1), N. I. Goreau (Fig. 6.8), R. C. Highsmith (Fig. 7.5), L. Kaufman (Fig. 8.6), R. F. G. Ormond (Fig. 9.2), G. W. Potts (Fig. 8.7) and J. Rubin (Fig. 3.2). And thanks are due to the many publishers and authors who gave their permission for the use of their figures, the sources of which are acknowledged in the legends.

The following photographs and line diagrams are my own work: Figs. 1.1, 1.5, 2.1, 2.4, 2.5*a,b*, 2.7*a*, 2.8, 2.11, 2.12*a,b*, 3.1*a,b,c*, 3.5, 3.6, 4.1, 4.2, 4.3, 4.5, 4.7, 5.1, 5.5, 5.6, 6.2, 6.3, 6.10, 7.2*a,b*, 7.3*b–e*, 7.8, 7.10, 7.11*a*, 7.12, 8.2, 8.4, 8.9, 9.1, 9.12, 10.1, 10.4 and are © G. F. Warner.

INTRODUCTION

This book is about the special contributions made to marine biology by the use of free diving as a research method. By free diving I mean the use of self contained underwater breathing apparatus (SCUBA) of which by far the most important form is the aqualung, pioneered in the 1940s by Jacques-Yves Cousteau, and consisting of a cylinder of compressed air, a demand valve and a harness. This relatively cheap and simple apparatus is easy to use after a period of training and permits far greater freedom of movement underwater than that allowed by traditional diving suits. It has provided marine biologists with the ability to explore the sea bed at first hand.

However, SCUBA diving has its limitations, the chief of which is depth. This causes problems in three ways. First, depth limits dive duration since the deeper one dives, the quicker one uses one's supply of compressed air. Second, nitrogen dissolves into the body under pressure; the deeper one dives and the longer one stays at depth, the more decompression time is required before surfacing to prevent nitrogen appearing as bubbles in the tissues and causing the 'bends'. Third, excessive nitrogen dissolved in the blood causes nitrogen narcosis or 'rapture of the deep'; the deeper one dives, the more one's senses are blunted. The effect becomes noticeable between 40 and 60 m and at 70 m most divers feel distinctly different; few people can make useful and safe scientific observations below this depth. To dive deeper safely and usefully requires a far greater investment in technology than most marine biologists (or their grant awarding bodies) can afford and I have therefore excluded from the scope of this book the fascinating discoveries made at much greater depths in the oceans by scientists using diving vehicles. These discoveries are, in any case, being made in rather different environments than those being explored by free divers. The realm of the free diver is the sublittoral: that part of the submerged coastline between low tide mark and about 60 m deep, delimited here by human diving physiology but also recognizable by its characteristic features as a major type of marine habitat.

The characteristic features of the sublittoral which distinguish it from other benthic marine habitats mostly derive from the proximity of the coastline and the sea surface. On the physical side, light, tidal currents and waves are almost universal features. Attachment to the substrate gives primary producers stability in the light and exposes organisms to relative water movement from the waves and currents as they swirl past. The biological characteristics of the sublittoral result from the physical features. Proximity to the coast augments the supply of plant nutrients, and primary production in both the attached and planktonic communities is usually relatively high. Water movement carrying food leads to the development of rich communities of sessile suspension-feeding animals which compete with each other for space on the substrate and interact with mobile predators. Similar communities to those of the sublittoral are to be found on the shore and on deep sea bottoms exposed to currents, but both these environments have obvious special features distinguishing them from the sublittoral.

Despite its characteristic features, however, the sublittoral is by no means a uniform habitat; there is a wealth of diversity resulting from both local and geographic conditions and also from the very nature of the characteristic features themselves. Water movement, light and nutrient levels are all variables; the first two in particular vary over quite small distances and depths. I have tried in this book to deal with this diversity both between and within chapters. To begin with, I have divided the book into four Parts, each dealing with a special type of sublittoral environment. The first Part, however, is used to introduce concepts which are developed further in later Parts. The development of these concepts through the course of the book is intended to bring out the essential character of the sublittoral. The Parts themselves are divided into chapters in which are discussed the responses of the living organisms to the various variables, and to each other. This last is an important area since modern research on community ecology is increasingly concerned with the interactions between community members in an attempt to refine hypotheses on such matters as species diversity, community dynamics and stability, and the extent to which communities are more than the sum of their parts. The sublittoral environment is particularly suitable for the study of these wider ecological issues because the life-spans of community members are relatively short and any changes therefore occur fairly quickly.

To return to diving, what makes the sublittoral the particular concern of divers rather than of some other type of marine biologist? The reason is that most sublittoral environments cannot be investigated in detail by any other method. Before the advent of SCUBA diving almost all knowledge of the sublittoral had to be inferred from what could be dredged, grabbed or grappled by boat from the sea bed, and these inferences supplemented by aquariaum studies and occasional glimpses by traditional diving or from submerged cameras. Dredging, etc. works reasonably well on soft substrates but very badly on rocky bottoms, and in no case is it possible to sample with any precision. SCUBA diving, however, works well on all types of substrate, especially rock, and divers are capable not only of precision sampling and collection of undamaged specimens, but also of direct observation and experimental manipulation of the environment. Imagine the contrast between a normal terrestrial ecologist and one attempting to work from a helicopter in a thick fog; diving has allowed sublittoral ecology to mature as a modern scientific discipline.

Part I: Hard Substrates

1

Physical factors and communities

1.1 Introduction

A rocky environment below low tide level is an exciting place for a diving biologist. The diversity of life is much greater than on a level substrate because of the wide variety of available habitats. There are smooth slopes, rugged ridges, cliffs, overhangs and caves, each of which supports a different assortment of sessile or sedentary organisms. This patchwork of communities is the foraging ground for a wealth of mobile animals ranging in size from protists, through tiny detritivorous gastropods and amphipods, to the predatory lobsters and fish. It is not easy to impose explanatory patterns on this riot of life but it is the business of this Part to attempt the task. One way to do so is to look for correlations with physical variables and for biological patterns of association and exclusion.

Many of the traditional physical variables of marine biology – temperature, salinity, oxygen tension, pH – are irrelevant as pattern formers on the small scale of, say, a rocky reef; they become important only when one is considering geographical distribution or large depth ranges. Light, however, changes sufficiently over small depth ranges, and with the topography of the rock, to affect markedly the local distribution of organisms. The biggest effect of light is to define the zone in which fixed primary producers – seaweeds, reef-building (hermatypic) corals – can grow. Below this zone the benthic communities are entirely heterotrophic, importing food from above in the form of displaced phytoplankton, other animals, bacteria and detritus.

Another important variable affecting life on rocky bottoms is water movement, including both wave action and currents. Many rocky substrates owe their origin and maintenance to the energy of water movement since it resuspends sediments that would otherwise bury them. Water movements negatively affect attached organisms by dragging at them and positively affect them by bringing nutrients and food and washing away excretory products. Patterns which relate to water movement are evident on rocky substrates and consist, on the one hand, of community differences based on the tolerances and requirements of community members, and on the other of adaptations at the individual or colony level to the prevailing type of water movement. By examining a community, therefore, a diver can often infer both the normal energy level of the environment and the usual direction of the current (assuming that this is not evident directly: most divers prefer to dive on calm days or at slack water so as to avoid the drag forces to which the organisms are adapted).

Linked in some respects with water movement is the variable of sedimentation. Water movements, if they are strong enough, wash sediment away. Thus high-energy environments are generally fairly clean whereas low-energy environments tend to accumulate silt and detritus during calm seasons of the year. Sedimentation, however, is not entirely dependent on the energy level; it also depends on the silt load of the local water. This varies greatly from place to place: coastal waters, especially near estuaries, generally have a heavy silt and detritus load whereas waters surrounding offshore or oceanic islands are usually cleaner. Sediment affects organisms by settling on them, necessitating either tolerance or a cleansing mechanism, and by providing food for deposit feeders. Potential sediment carried in currents affects organisms by bombarding them.

Biological variables which affect the composition of communities include the productivity of the local water and the interactions between community members. The productivity, or biological energy content, of the water is important since the majority of marine primary consumers are suspension or deposit feeders, feeding on plankton and detritus rather than on seaweeds. Interactions between community members include trophic interactions such as predation and grazing, competition for space, and various types of commensalism.

This Part is mainly concerned with environments below the zone inhabited by fixed primary producers. This is because sublittoral seaweeds (mostly kelps) and hermatypic corals so dominate their environments that they impose ecological patterns which are intimately connected with their own biology. They therefore deserve special treatment and form the subjects of Parts II and III. This is not to say that the contents of this Part are irrelevant to the study of kelp forests and coral reefs – quite the reverse. All the processes dealt with here occur importantly in kelp- and coral-dominated habitats; indeed, they can be regarded as general principles. The main questions addressed in this Part are: Chapter 1, in what respects do the relevant physical variables – light, water movement and sedimentation – affect the nature of communities? Chapter 2, in what ways do the sedentary and sessile animals tolerate and exploit water movement? Chapter 3, how do the members of communities interact with each other? All these questions recur in one form or another in subsequent chapters; this is partly because they relate to general principles, but also because in some cases adequate study has only been carried out in other environments. This is especially true in the case of biological interactions involving mobile animals such as fish where the majority of the work has been carried out on coral reefs (see Chapter 9).

1.2   The communities

Below the level of the kelp forests or coral reefs, or within these habitats in dimly lit places such as caves, overhangs and cliffs, the communities are dominated by sessile, usually colonial animals. These comprise various coelenterates (hydroids, octocorals, anemones, ahermatypic corals), sponges, bryozoans and ascidians (Fig. 1.1). Usually present, sometimes in very large numbers, are serpulid worms, barnacles and sessile molluscs such as oysters and mussels. The larger sessile animals, especially the sponges, octocorals, anemones, large hydroids and large arborescent bryozoans, form the 'trees' and 'thickets' of this miniature animal forest. The 'undergrowth' is composed of smaller sessile animals which may also grow as epizoites on the surfaces of the larger animals. Amongst this animal 'vegetation' lives a wealth of small, mobile species such as amphipods, isopods, polychaetes, gastropods and brittle stars, and an occasional larger animal such as a crab or starfish. Fish cruise above the surface and around the sponges and octocorals.

The primary food source of this community is the plankton and detritus carried in the water. Thus suspension feeding is the most important method of primary consumption. Sponges, bryozoans, ascidians, serpulid worms, barnacles and mussels are all suspension feeders, probably mostly ingesting phytoplankton and detritus. The colonial coelenterates are commonly regarded as carnivorous suspension feeders, taking zooplankton, but some evidence of phytoplankton ingestion has been obtained for octocorals. Some anemones are also suspension feeders (e.g. *Metridium senile*, Fig. 2.9) but most are predators which lie in wait and snare their prey – an incautious fish or shrimp. Many of the smaller mobile animals are suspension feeders – some of the amphipods and brittle stars – but most are deposit feeders, benefiting from the deposition of detritus and probably also from the faeces and mucus cleansing mechanisms of the larger animals. There are also some small predators such as nudibranchs and pycnogonids, feeding on the coelenterates, sponges and bryozoans, and aphroditid polychaetes

Fig. 1.1. The sessile community on a vertical rock face at 20 m off Portland Bill, English Channel. White anemones *Actinothoe sphyrodeta* and fleshy octocorals *Alcyonium digitatum* are prominent. In the centre is a colony of the branched bryozoan *Cellaria fistulosa*, and sponges and ascidians are also present.

feeding on the small deposit feeders. Most of the larger animals are predators (crabs, starfish, sea-urchins, fish) but some are suspension feeders (crinoids, some holothurians) and some are deposit feeders (other holothurians). It may sound strange to describe sea-urchins as predators, grazing is a better description of their feeding method; in this environment, however, except for occasional detritus, their grazed food is entirely animal.

1.3   The effects of light

The boundary between suspension-feeding communities and those dominated by primary producers is not sharp: the algae or reef corals gradually decrease in density as the depth increases. Some of the sessile animals that occur commonly in the deeper, darker communities are also present in the shallower, lighter regions living alongside the primary producers. Many, however, are restricted in this environment to the undersides of rocks and to other relatively dark places and only deeper down are they found growing out in the open. This habit may be a direct response to light or it may be that these animals cannot successfully compete for space with algae. The growth rates of algae are usually much greater than those of sessile animals and experimental illumination of sessile communities often results in their rapid overgrowth by algae. The dense growth of algae leads to extra sedimentation and this may kill the animals if algal overgrowth has not already done so. It is thus extremely difficult to separate the effect of light on an animal species from the effect of light on the community as a whole.

Direct effects of light on the short-term behaviour of animals are easier to demonstrate. Thus the jewel anemone *Corynactis* expands in dim light and at night but contracts the tentacles and closes when exposed to daylight. The same is true of many corals, including most reef-builders (§ 7.2), but *Corynactis*, unlike the reef corals, does not occur in places likely to be exposed to direct light. Animals such as the suspension feeding basket stars (Fig. 2.6) unfurl their branched arms at night, but I have observed them to extend an arm or two by day when the light intensity was low due to depth or to turbid water.

1.4 The effects of water movement

Adaptations at the individual and species level to the type and degree of water movement are described in Chapter 2. Here I will consider the effects of the ambient energy level of the environment on the nature of the community as a whole. Before proceeding, however, the term ambient energy level needs some explanation. The flows actually experienced by a given animal community may vary enormously. They vary with the seasons through such agencies as winter storms, they vary from hour to hour through the ebb, slack and flow of the tide, they vary from minute to minute through the turbulence of the water flow close to the substrate, and they vary with proximity to the bottom through frictional drag: flow is slowest at the bottom and increases to 'mainstream' some 30–100 cm above it. Undoubtedly in some environments the average mainstream flow rate is a meaningful way of indicating the energy level, but in others the maximum flow rate, the frequency of storms, or the duration of slack water, may be of overriding importance. Average flow rates are, in any case, underestimates; maximum flow rates are hardly ever measured since diving is too dangerous at such times.

Violent water movements (> 1 m s^{-1}), especially wave surge, generally have negative effects on communities. The most extreme cases occur when sand or pebbles are carried in the surge and the consequent scouring leads to the complete absence of life. This situation is common in shallow water (0–10 m) at the bases of cliffs and along the floors of gullies and caves. In deeper water, currents can produce a similar effect where the bases of rocks meet a sandy substrate. An analogous lifeless zone has been found in the top 5–15 m of the sublittoral of rocky coasts in the arctic and antarctic and is due to scouring by winter pack ice which moves up and down with the tides (Fig. 1.2). In the absence of abrasion, violent wave action leads to the development of a low diversity community characterized by strong, tough or streamlined organisms adapted to tolerate the fierce pounding of the waves. Since the water movement produced by wave action decreases rapidly with depth these communities generally occur in shallow water and, because of the high energy of the environment, are rarely studied by divers.

Moderate water movements (10–100 cm s^{-1}) produced by tidal

Fig. 1.2. Vertical zonation at McMurdo Sound, Antarctica. Zone I, scoured by pack ice, contains a few mobile animals such as sea urchins, starfish, nemertines and pycnogonids. Zone II, less physically disturbed, is dominated by sessile coelenterates: anemones, hydroids and fleshy octocorals. Zone III, least disturbed, is dominated by sponges. Reproduced with permission from Dayton *et al.* (1970). Copyright: Academic Press Inc. (London) Ltd.

Zone I 0–15 m

Zone II 15–30 m

Zone III below 33 m

or ocean currents, or by waves passing 5–15 m overhead, usually have positive effects on communities (Fig. 1.3). This is because water movements bring food in the form of plankton and detritus to the suspension feeders. In a survey of a rocky hogs-back reef off southern California, Pequegnat (1964) found that the top of the reef, 10.7 m below the surface and experiencing mean wave-generated flow rates of 34 cm s^{-1}, supported very dense populations of suspension feeders. Almost all the available space was colonized by sessile animals. Rock oysters, *Chama pellucida* occurred stacked one on top of another in crusts up to 40 cm thick in which the highest population density was more than 4000 m^{-2}. These crusts provided a habitat for over 100 other animal species leading to community figures of up to 35000 individual organisms per square metre. Elsewhere on the reef-top, sponge mats 8 cm thick were dominant, sometimes associated with bryozoans and hydroids, or with dense populations (30000 m^{-2}) of the small

Fig. 1.3. The wall of a bluehole (marine cave) in Andros Island, Bahamas. Hydroids up to 75 cm long stream in a tidal current of 30–50 cm sec^{-1}; the rock is thickly encrusted with non-symbiotic corals and sponges. From Warner & Moore (1984).

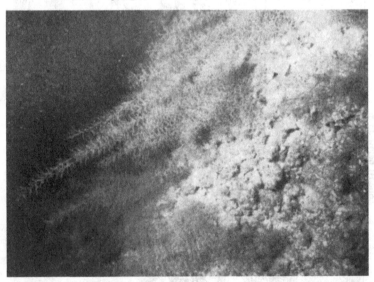

bivalve *Hiatella arctica*. On the more turbulent parts of the reef crest the oyster crusts and sponge mats were replaced by aggregations of vermetid gastropods, up to 650 m^{-2}, or by thickets of tall gorgonians, 15 m^{-2}, or patches of jewel anemones, 3000 m^{-2}, or carpets of colonial ascidians. On the steep walls of the reef, 13.7 m below the surface and with a mean flow rate of about 13 cm s^{-1}, the community was less rich, bryozoans, sponges and colonial ascidians were dominant and often formed extensive mats or carpets, but these were not as thick as on the reef crest. The base of the reef, at 17 m deep and with a mean flow rate of 7 cm s^{-1}, had the poorest sessile fauna composed of thin crusts (2–3 mm) of bryozoan and sponge; however, sea urchins and deposit-feeding holothurians were quite abundant. Around the base of the reef on the surrounding sediment scavengers and deposit feeders were common and no doubt mostly fed on the rain of material dislodged from the reef above.

In communities subject to slight water movement (0–5 cm s^{-1}) diversity and biomass are usually reduced. Free space on the rock surface, often liberally coated with sediment, is common. Deposit feeders such as holothurians and terebellid worms become more important although suspension feeders may still dominate. The absence of water movement and the deposition of sediment combine to make this a difficult environment for sessile animals.

1.4.1 *Passive and active suspension feeders*

Many suspension feeders simply stretch a food-catching net across the direction of water movement and filter particles from suspension. These are called passive suspension feeders and include colonial coelenterates such as hydroids and gorgonians, catching zooplankton, and echinoderms such as crinoids, catching phytoplankton and detritus. Clearly, for these animals, the delivery rate of food is proportional to the current speed, provided only that their feeding efficiency can be maintained through the environmental range of flow rates. In contrast, animals such as sponges, bryozoans, ascidians, serpulid worms and bivalve molluscs are active suspension feeders and use their own energy to drive water through their feeding structures. Active suspension feeders benefit from water movements since the currents constantly renew their source of food and help to keep fine particles in suspension. There is evidence (e.g. Fig. 2.10) that active suspension feeders can use the environmental current to augment their own feeding currents, but at low current speeds their feeding rate is independent of flow rate. One might expect, therefore, that an important effect of water movement on communities would be to influence the proportions of active and passive suspension feeders: with decreasing water movement the proportion of passive suspension feeders should drop. Various studies have confirmed this trend. The difference was apparent in a comparison between the macrofauna of the sublittoral of the open coast of North Wales (vigorous water movement) and that of an almost completely enclosed quarry in the same area (water circulated by tides but water movement adjacent to the epifauna practically nil) (Table 1.1). The diversity of the community within the quarry was found to be lower than that of the open coast; active suspension feeders, although still dominant, comprised fewer, and different, species than outside the quarry. Some of the active suspension feeders which were common within the quarry were rare or absent outside, indicating special adaptations to reduced water movement and probably also to increased sedimentation.

Table 1.1 *Distribution of feeding types of prominent species on the open coast and in a sheltered quarry in North Wales*

Feeding group	Open coast only or predominantly open coast		Quarry only or predominantly quarry	
	number of species	% of species	number of species	%of species
Passive suspension feeders	7	17.5	1	5
Active suspension feeders	18	45.0	14	61
Passive carnivores	5	12.5	3	12
Active carnivores	3	7.5	0	0
Grazers	1	2.5	1	5
Omnivores	5	12.5	3	12
Deposit feeders	1	2.5	1	5
Total	40		23	

After Hiscock & Hoare (1975).

Another study, on the epizoites of the large sublittoral hydroid *Nemertesia antennina*, showed the same effect on a much smaller scale (Hughes, 1975). This hydroid grows in clumps 15–30 cm high in places exposed to tidal currents. Flow measurements next to the hydroids at 3, 9 and 15 cm above the bottom demonstrated a gradient in current velocity with mean speeds of the order of 2–5 cm s^{-1} at the bottom increasing to 7–15 cm s^{-1} at 15 cm. Epizoites on the distal parts of the host exposed to the faster currents, were mostly passive suspension feeders and included hydroids and tubicolous amphipods (see Fig. 3.6). On the proximal parts of the host the commonest organisms were active suspension feeders (bryozoans, ascidians, bivalve molluscs, sponges) and deposit feeders (amphipods, gastropod molluscs).

1.5   The effects of turbidity and sedimentation

The effects of these two variables are intimately linked both to each other and to light and water movement. First, high turbidity (e.g. 10–500 mg dry weight suspended particulate matter per litre) caused by silt, detritus or plankton, reduces the penetration of light through the water and consequently raises the depth limit of the fixed primary producers. Second, the balance between turbidity and sedimentation is, to a great extent, governed by water movement: currents and waves, by resuspending sediment or by keeping silt in suspension, tend to increase turbidity; conversely, calm seasons or periods of reduced current lead to sedimentation in the environment and to decreased turbidity.

Laboratory and field experiments show that turbidity and sedimentation generally have deleterious effect on animals (Moore, 1977). This is especially true of suspension feeders in which turbidity leads to the cleansing and sorting mechanisms being overworked, to a consequent drop in filtering rate, and a reduction in growth rate (Fig. 1.4). Bivalve molluscs, with increasing turbidity, first increase their rate of pseudofaecal production, then decrease their filtering rate. Sponges decrease their filtering rate partly because the inhalant pores become blocked with particles. The seriousness of this curtailment of feeding depends, to some extent, on the nature of the particles. Turbid water usually contains many inorganic particles; these may have some nutritive value due to adsorbed bacteria but their food value is much less than that of plankton or detritus. To survive in turbid conditions suspension feeders need adaptations to cleansing and sorting mechanisms which will permit adequate feeding. These have been achieved in species from most groups and thus the main differences between

Fig. 1.4. Effects of turbidity on the slipper limpet *Crepidula fornicata* showing (*a*) that growth is slower close to the bottom (higher turbidity) and in areas of overall high turbidity, and (*b*) that one explanation for this may be the effect of turbidity on the rate of active suspension feeding. Points are means ± 2 standard errors. Modified from Johnson (1972).

(*a*)

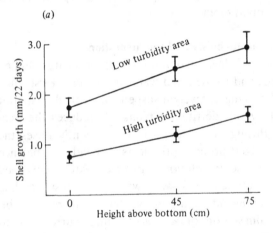

(*b*)

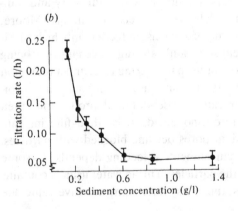

turbid and clear water communities are species differences. Communities still contain sponges, bryozoans, coelenterates, ascidians, etc., but the species are different from those in clear environments and the community diversity is generally lower. If conditions are very turbid, however, the biomass per unit area may be severely reduced.

When turbid water occurs in a low energy environment, or when a rough sea calms down, the potential sediment in the water – silt, detritus, faecal pellets – settles out on the bottom. This sediment settles on top of animals (Fig. 1.5) which, if they are to survive, must be able to unbury or cleanse themselves. As in the case of turbid environments, some species from most groups have these capabilities, but communities living in areas of high sedimentation tend to have lower diversity and lower biomass than those in clean areas. Hydroids appear to be particularly susceptible to sedimentation but this may be because of their dependence, as passive suspension feeders, on water movement. Encrusting animals, such as sheet-like bryozoans, are susceptible to sedimentation because

Fig. 1.5. Colony of the fleshy octocoral *Alcyonium digitatum* amongst the lower branches of which a pool of sediment has collected. The polyps of the octocoral are retracted; the brittle star is *Ophiocomina nigra*. 7 m, Millport, Scotland.

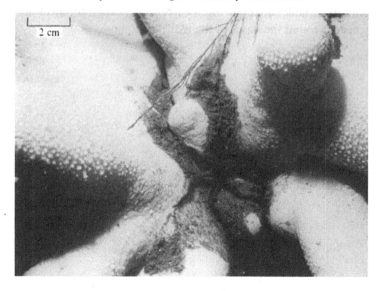

2 cm

of their shape; in environments liable to sedimentation they tend to occur on steep slopes or overhangs where sediment is unlikely to remain without falling off. The only primary consumers which clearly benefit in these environments are deposit feeders.

In his review of the effects on organisms of turbidity and sedimentation, Moore (1977) concluded that this was a neglected field but that 'perhaps the problem has been brought into sharper focus today with the divers' overwhelming preoccupation with underwater visibility'. Although this is probably true, it is also true that divers, for obvious reasons, tend to avoid extremely turbid conditions, particularly when these are caused by rough weather, and so lack knowledge of them. Yet in variable environments the extremes may be very important. Hughes (1975), during his work on the hydroid *Nemertesia*, measured potential sediment 0–4 cm above the bottom by means of small sediment traps. These traps caught ten times more sediment on a rough day than had been caught per day on five previous calm days. The greatest weight of sediment on the rough day was caught in the 0–1 cm layer indicating that the extra turbidity was caused by resuspension. Investigation of the weather records showed that rough days were quite common, especially during the winter, and Hughes concluded that bombardment by particles and sedimentation were important variables affecting the occurrence and distribution of the organisms around the bases of the hydroid clumps.

2

Adaptations of organisms to water movement

2.1 Introduction

One of the clearest patterns visible to a diver is the orientation of fan-shaped passive suspension feeders to currents and wave surge. A large number of different species dispose their food catching surfaces in the form of a fan orientated at right angles to the direction of the current (Fig. 2.1). The best known are the gorgonians, but other colonial coelenterates – hydroids, black corals – and various echinoderms, especially crinoids and basket stars, adopt the same posture. There is an extensive literature on the subject all of which depends, at least for initial observations, on

Fig. 2.1. Orientation to wave action in sea-fans *Gorgonia* sp. which are bending from right to left as a wave passes overhead. 4 m, Tobago, Caribbean. The large fan in the centre is about 1 m high.

diving studies. These studies have shown an intimate link between the type of water movement and the shape of the suspension feeder – which need not be that of a fan. They have also led to a dynamic consideration of the organisms in which there is often a partial conflict between the need to avoid being damaged by moving water and the need to exploit it as a source of food.

2.2 Orientation in passive suspension feeders

Research on this topic began on coral reefs with gorgonians orientated to the prevailing bidirectional wave surge. An early finding was that orientation improved with size: large fans were all found to be orientated at right angles to the surge but small fans were randomly orientated (Wainwright & Dillon, 1969). This finding was explained by postulating the existence of a turbulent, near-bottom layer in which flow was multidirectional (the existence of such a layer has since been demonstrated); small fans inhabiting this layer lack orientational cues and are therefore orientated at random. It was convincingly argued that, for fan-shaped animals, an orientation at right angles to the prevailing surge was hydrodynamically the most stable (Fig. 2.2) and it was suggested that as

Fig. 2.2. Diagram, as viewed from above, of the effects of current on various sea-fan orientations; arrows indicate current direction and relative magnitude of drag forces, current speed is equal in all cases, dotted lines indicate orientations in the absence of a current. In (*a*) the fan is parallel to the current and under ideal conditions both drag and twist are minimal, but this orientation is very unstable since a slight alteration in current leads to condition (*b*). In (*b*) the drag forces on the two sides of the fan are very unequal and deliver maximum twist to the stem. In (*c*) there is more total drag but much less twist, and in (*d*) no twist but maximum drag. (*d*) is the stable orientation since slight alterations in current direction cause only slight alterations in forces. Modified from Wainwright & Dillon (1969).

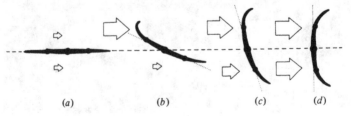

(*a*)   (*b*)   (*c*)   (*d*)

the small fans grew out of the turbulent near-bottom layer into the bidirectional mainstream surge, they would be passively twisted by the surge into this stable orientation. Stems of large fans were sectioned and showed evidence of having been twisted and of the twist having been permanently fixed by subsequent growth (Fig. 2.3). This argument provides a mechanical explanation of orientation but does not explain the occurrence of a fan shape in the first place: a radially symmetrical bush would be just as stable and would not be subject to twisting in response to a change in current direction. In fact, many gorgonians, hydroids, etc. do have bush-like growth forms and some species vary their growth form according to their environment. Attention was therefore focused on species in which the growth form – fan or bush – varied in apparent relation to the type of water movement. It was found that the more directional the current or surge became, the more the form of the colony flattened into a fan; conversely, bushy shapes were found to be associated with turbulent environments (e.g. Velimirov, 1973). These findings led to a trophic hypothesis to explain sea-fan shape: that a fan shape orientated across the current is the most efficient filter-shape for passive suspension feeding in directional currents; in turbulent environments a bush-shape is more efficient since it can exploit currents from all directions. Partial support for this hypothesis has been obtained by a laboratory study in which a fan-shaped gorgonian caught more food when orientated across the current

Fig. 2.3. Cross-section of the stem of a sea-fan *Gorgonia* sp. The dashed line shows the orientation of the fan when collected and the solid line shows a possible earlier orientation, suggesting that the stem has twisted during growth. From Wainwright & Dillon (1969).

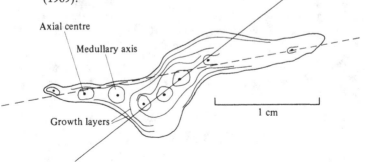

Axial centre

Medullary axis

Growth layers

1 cm

than when orientated edge-on to the current (Leversee, 1976). A final point in favour of developing a fan-shape is that, if there is relative movement between branches, a fan-shape minimizes the chances of self-abrasion. Most of the best developed gorgonian sea-fans are not only quite flat, they also show complete anastomosis which removes the possibility of relative movement between branches (Fig. 2.4).

Mobile passive suspension feeders, such as crinoids, complement the patterns seen in sea-fans and provide support for the trophic explanation of fan shapes. Crinoids feeding in directional currents or in bidirectional surge dispose their arms to form a filtration fan at right angles to the current; at slack water or in turbulent environments where the flow is multidirectional the arms are held in a radial arrangement. This behavioural difference also extends to the pinnules of the arms which, in directional flow, stand in a comb-like row on either side of the arm, but in multidirectional flow stand out of alignment with each other to form four or more radially arranged rows along the length of the arm (Meyer, 1982) (Fig. 2.5). Similar behavioural changes related to the type of water

Fig. 2.4. Portion of the blade of a sea-fan *Gorgonia* sp. showing the tight reticulation achieved through anastomosis. Note expanded polyps. 10 m, Jamaica.

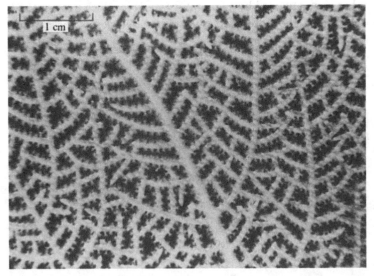

1 cm

Fig. 2.5. Passive suspension feeding in crinoids. The photographs show the arms of feeding crinoids in: (*a*) a directional current, the pinnules and tube feet are arranged in a single plane at right angles to the flow, and (*b*) in turbulent water, the pinnules are arranged in two planes at right angles to each other so as to collect food approaching from any direction.

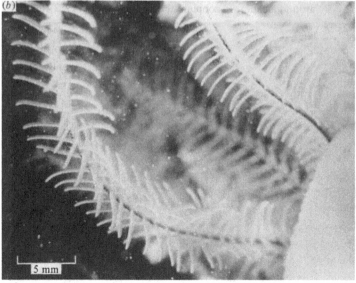

movement have been observed in the tube feet of suspension feeding brittle stars (Warner, 1982). In all cases the radial arrangements are analogous to the bush-shapes of sea-fans – they appear to be adaptations to exploit the food contained in currents coming from any direction.

So far I have considered orientations to tidal (reversing) currents and to wave surge and have contrasted these with arrangements adapted to turbulent flow. Some environments, however, are exposed to persistent unidirectional (non-reversing) currents either through oceanic circulation or through local circumstances. Here, instead of the fan being flat it develops the shape of a parabolic dish with the concave side facing the current. This orientation has been recorded in various gorgonians, in several species of black coral and in some stalked crinoids. It has also been observed in some mobile passive suspension feeders and one of the most spectacular examples is that of the basket stars which unfurl their branched arms at night to catch zooplankton (Warner, 1982) (Fig. 2.6). Laboratory experiments with models indicate that a parabolic

Fig. 2.6. A basket star at the top of a sea-fan, photographed at night at about 4 m, Trinidad, Caribbean. The feeding arms form a concave dish. Arrow shows a knot of feeding branchlets folded around a captured zooplankter. From Warner (1982).

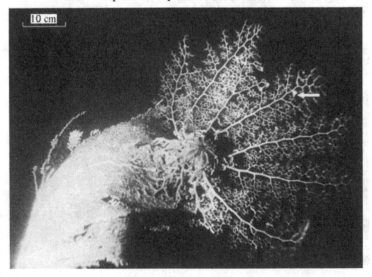

dish-shaped filtration fan may filter more water than a flat fan of the same surface area, thus the advantage of these parabolic fans probably lies in an increased filtering efficiency (Warner, 1977). An interesting observation on parabolic fans is that, in many cases, the food collecting organs are situated on the convex side of the fan, to leeward in relation to the current. This occurs in some dish-shaped gorgonians and black corals in which the polyps, which capture zooplankton, are borne predominantly on the leeward side of the fan (Fig. 2.7). The same phenomenon occurs in crinoids and basket-stars in which the arms are orientated so that the aboral side faces the current. This leeward placement of food catching surfaces is probably an adaptation to the small-scale flow patterns around the branches which make up the filtration fan (Fig 2.8). The turbulence to leeward of the branches provides a microenvironment of relatively low current speed where particles swirl round two or three times before drifting on their way. It seems likely that food capture is more efficient within these leeward eddies than elsewhere on the fan (Lasker, 1981).

2.3   Shape and stiffness in coelenterates

It is argued above (§ 1.4.1) that as current speed increases, so too does the proportion of passive suspension feeders in the community. However, passive suspension feeders occur over a wide range of environmental energy levels. Many species are characteristic of particular energy levels, but some can adapt by means of suitable growth forms to a range of energy levels. A diving study in the Isles of Scilly (Robins, 1968) on two octocorals, *Alcyonium digitatum* and *A. glomeratum*, which appeared to inhabit high and lower energy environments respectively, indicated that the shape and stiffness of the colonies were adapted to the level of water movement. Flow in the environments of the two octocorals was mostly caused by surge and the to and fro movement of suspended particles was measured underwater. *A. digitatum* was found to be exposed to surge of 30–100 cm whereas *A. glomeratum* only had to contend with surge of 0–30 cm. *A. digitatum* forms squat fleshy colonies with few short branches (Fig. 1.1) whereas *A. glomeratum* forms graceful dendritic colonies with thinner, more numerous branches. Examination of the skeletal

Fig. 2.7. The fan-shaped black coral *Antipathes atlantica*, 25 m, Trinidad. (*a*) diver recording the orientation of a dish-shaped colony with concave side facing the current. (*b*) close-up of the convex side of the dish showing all polyps facing downcurrent.

system revealed a better developed hydrostatic and spicular skeleton in *A. digitatum* than in *A. glomeratum*. The well developed skeletal system and short thick branches enable *A. digitatum* to stand up to the high drag forces of its environment whereas the dendritic growth form of *A. glomeratum* enables it to extend through, and to feed in, a relatively large volume of its calmer environment. Similar differences have been observed between colonies of fan-shaped gorgonian species inhabiting different energy level environments: a growth form with fewer thicker branches is characteristic of environments with vigorous water movements whereas delicate more frequently branched colonies are found in calmer environments.

Similar adaptations were found to apply to two anemone species studied in the intertidal and sublittoral off the coast of Washington (Koehl, 1977 a, b, c). *Metridium senile* is a tall (38 cm) anemone occurring in the sublittoral and exposed to steady tidal currents of about 10 cm s^{-1}; it is a passive suspension feeder on zooplankton. *Anthopleura xanthogrammica* is a short (2.5 cm) broad (6 cm) anemone occurring in large numbers on the bottoms of intertidal surge channels on exposed rocky coasts. Mainstream surge in the channels at high tide is more than 2 m s^{-1} and the anemones feed by seizing mussels which fall from the rocks above after having been dislodged by the waves. The shapes and stiffnesses of the two anemones are adapted to their habitats and ways of life. *M. senile*,

Fig. 2.8. Diagrammatic cross-section through a branch of a passive suspension feeder in a current showing the streamlines becoming turbulent on the leeward side.

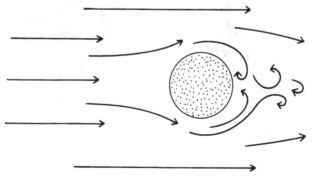

being tall and having a thin flexible body wall, extends into mainstream currents and is bent into the posture shown in Fig. 2.9. This posture, with the broad lobed oral disc orientated at right angles to the current and the tentacles to leeward, is ideal for its suspension feeding habit. In contrast, the squat shape of *A. xanthogrammica* ensures that it is not exposed to the violent mainstream velocities of the surge channels. Flow just above the oral discs was usually about 1 m s^{-1} and was more turbulent than the mainstream surge. As a further consequence of its shape – most of its surface area being parallel to the flow – and having a stiffer thicker body wall than *M. senile*, *A. xanthogrammica* is not bent by the current but remains facing upwards, awaiting its prey.

Flexibility in sea-fans is an interesting topic: differences in stiffness between species seem to be related to the flow regime and also possibly to the mode of feeding. Gorgonian sea-fans on coral reefs, such as *Gorgonia ventalina*, wave back and forth as the waves pass overhead (Fig. 2.1). They appear very flexible when contrasted with species living in deeper water and exposed to tidal or oceanic currents. These deeper water species may be bent by strong currents but for most of the time they stand up to the flow, which filters between the branches. One way in which stiffness is altered is by varying the cross-sectional shape of the stems and branches.

Fig. 2.9. The anemone *Metridium senile* bent over by a tidal current. (*a*) side view, arrow indicates current direction; (*b*) view from upcurrent. From Koehl (1976).

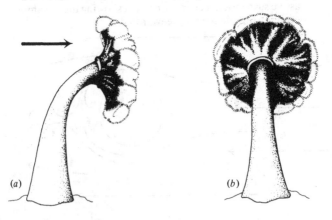

G. ventalina stems and branches are elliptical in cross-section with the long axis of the ellipse parallel to the plane of the fan (Fig. 2.3), facilitating bending with the waves. Many deeper water gorgonians and fan shaped black corals also have elliptical stems and branches but here the long axis of the ellipse is at right angles to the plane of the fan and increases resistance to bending (Wainwright *et al.*, 1976; Warner, 1981). Nutritionally it is important for the deeper water species to stand up to the flow since the filtering efficiency of the colony depends on the plane of the fan being at right angles to the current. Most shallow-water coral reef gorgonians, however, contain symbiotic unicellular algae in their tissues which may, through photosynthesis, provide a supplement to the normal diet of filtered plankton. The nutritional requirement for shallow-water fans to stand up to currents may, therefore, be less than that of the deeper water species. Additionally, living in shallow water, the sea-fans are exposed to the full vigour of storm waves and probably require flexibility to bend out of trouble when the sea is rough. To bend with the current reduces the likelihood of damage in three ways: (1) during bending the relative movement of water past the branches is reduced, (2) the drag on a bent colony is reduced since most of its surface lies parallel to the flow, (3) as the colony bends it moves out of the mainstream into the less rapidly moving layers of water close to the substrate.

Alteration of the cross-sectional shape of the stem and branches is not the only way to alter flexibility. The skeletal systems of sessile coelenterates show a great deal of variety, much of which may be related to adaptational variation in stiffness. Fleshy octocorals and anemones have hydrostatic skeletons which provide a range of stiffnesses such as that between *Metridium senile* (flexible) and *Alcyonium digitatum* (stiff); in *A. digitatum* the body wall is further stiffened by reinforcement with spicules. An advantage of the coelenterate hydrostatic skeleton is that, by expelling water, the body volume can be reduced allowing the organism to contract close to the substrate in rough weather. Both *M. senile* and *A. glomeratum* are highly contractile. In gorgonians and black corals contraction and shape alteration are prevented by a horny axial skeleton. In gorgonians the axial skeleton may be reinforced by calcification, or the coenenchyme surrounding the axis may be stiffened with spicules. A very wide range of possible shapes and stiffnesses is therefore available to these sessile organisms and the functional significance of the different systems found in different species is only beginning to be understood (Muzik & Wainwright, 1977; Koehl, 1982).

2.4 Active suspension feeders

Water movement stimulates the growth of active suspension feeders by constantly renewing their source of food. Since they create their own feeding currents their minimum feeding rate should be independent of environmental flow rate. By appropriate orientation, however, they should be able to augment their own currents by tapping the energy of the environmental flow. Little work has so far been carried out on this aspect of active suspension feeding. Adaptive shape in sponges has, however, received some attention and there are occasional records of orientation to currents in other groups.

Suspension feeders such as barnacles, which actively beat a dish-shaped cirral net through still water, become almost passive in a

Fig. 2.10. The effect of environmental flow on the pumping rate, or outward flow through an osculum, of the sponge *Halichondria*. Closed circles – live active sponge; open circles – sponge inactivated by exposure to fresh water. From Vogel (1974).

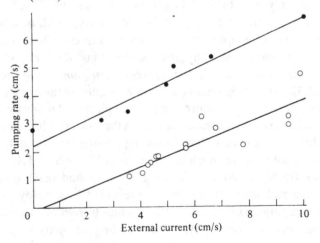

current by turning their nets to face the flow. I have observed similar behaviour in porcelain crabs which carry dish-shaped nets on each third maxilliped. Turning the branchial syphon into the current has been observed in some ascidians (Young & Braithwaite, 1980) and it should be possible for bivalve molluscs to orientate themselves so that the inhalant syphon faces the current. This arrangement has been observed in scallops which, in a unidirectional current on a level sandy bottom, were found to be facing the current, using it to augment their own ciliary flow (Hartnoll, 1967).

In sponges, most work has focused on the position and shape of the oscula or exhalant apertures. The oscula are normally borne on the tips of volcano-shaped prominences rising above the surface of the sponge. Their relative elevation results in their being exposed to slightly faster moving water than are the numerous tiny ostia or inhalant pores, which are scattered over the surface. The difference in current speed results in a passively induced flow from ostia to oscula. Various physical principles are responsible for the

Fig. 2.11. Colony of the encrusting sponge *Halichondria* in which the surface has grown to form parallel ridges with oscula along the crests. The ridges were at right angles to the current. 8 m, the Needles, English Channel.

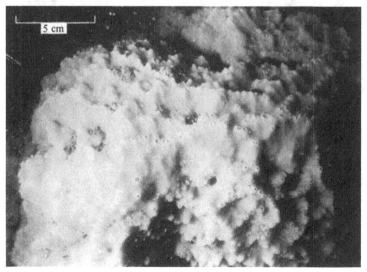

Fig. 2.12. Colony of the sponge *Xestospongia* forming a single ridge with marginal oscula, at right angles to the current. (*a*) side view; (*b*) view from above. The sponge was about 70 cm long. Note several species of bushy gorgonians (Octocorallia) on the surrounding reef-rock. 10 m, Tobago, Caribbean.

induced current and the result is similar to the updraught in a chimney (Vogel, 1978). Observations on living sponges have shown that the rate of flow through the oscula is positively correlated with the environmental flow. This is also true of freshly killed sponges, but the oscular flow rate of a dead sponge is a standard amount less, at any given environmental flow rate, than that of a living sponge. The deficit represents the active pumping of the living sponge (Fig. 2.10). Sponges are notable for their plasticity of form and oscular heights may vary within species between quite wide limits. In general, where environmental currents are slight, oscula occur on tall chimneys; where currents are swift they occur on short volcanoes. Presumably these growth forms are adaptations to augment the active pumping rate by a standard amount to achieve a fairly constant, species-specific, internal flow rate. A common arrangement in regular reversing currents (tidal or wave-generated) is for the sponge to grow in the form of a ridge, or, in an encrusting sponge, a series of ridges, with the oscula arranged in a row along the crest of each ridge at right angles to the current (Figs. 2.11 & 2.12). A single ridge may develop into a fan-shape with marginal oscula exposed to the faster current past the edges of the fan. Branching sponges in environments with directional currents may also grow to form fans orientated across the flow, here the oscula often open into the spaces between branches where the current through the fan is swiftest.

3

Biological interactions

3.1 Introduction

So far I have considered the effects of physical variables, especially water movement, on the nature of communities. However, communities are more than just a collection of organisms which live together merely because they can all tolerate the same environmental conditions. Increasingly, over the last few years, biological interactions have been shown to be important in determining not just the detailed organization of communities but often their fundamental nature. Many of these studies have not involved diving; intertidal work and studies on the development of communities on artificial panels have made a substantial contribution. However, diving has been important in improving awareness of biological interactions and has provided some spectacular examples such as the effects of urchins on kelp forests (§ 5.5) and of the crown-of-thorns starfish on coral reefs (§ 8.4). Diving work has also been important as a cross-check on panel studies since panels, although very useful for experimental purposes, are unlike natural substrates in a number of respects. Not least of these is the normal isolation of panels from established communities, leading to colonization of panels from the plankton only; a bare area occurring in an established community is colonized in addition by benthic larvae and by growth of colonies around the periphery.

Three sorts of interaction are considered in this chapter: trophic, competitive and commensal. The trophic interactions mostly concern slow moving or sessile predators like starfish and sea anemones; faster moving predators have been observed most frequently on coral reefs and are considered in Chapter 9. The competitive interactions described here involve competition for space and lead to a discussion of community dynamics and succession. Commensal interactions include positive associations between organisms other than predator/prey or host/parasite relationships; most concern one species exploiting another for living space or for protection. Mutual exploitation, or symbiosis, is commonest on coral reefs and examples of symbiotic behaviour are given in chapters 8 and 9.

3.2 Trophic interactions

Predators probably have very big effects on the nature of sessile animal communities. These effects may either constitute a continuous pressure on the community or may be intermittent. An example of continuous predator pressure is the effect of grazing by sea urchins. One population of 20–100 million *Echinus esculentus* were found off the coast of Helgoland, at a density of about $7 \, m^{-2}$, feeding on a rock-boring polychaete worm *Polydora ciliata* by biting into the rock; it was estimated that this activity was eroding the rock at a rate of 30000 tons per year (Krumbein & Pers, 1974). Heavy urchin grazing has profound effects on the sessile fauna and often leads to a general decrease in biomass and to the creation of relatively barren grounds. Gulliksen *et al.* (1980) described a diving study on an arctic island where, eight years previously, a volcanic eruption had produced an extensive new underwater larval platform. The divers compared the fauna of the new grounds with that of adjacent old larval grounds. Above 15 m there was an ice-scoured zone on both new and old grounds (cf. Fig. 1.2) but below 20 m considerable differences were apparent. They found a high biomass on the new grounds, due largely to a dense population of the suspension feeding bivalve *Hiatella arctica*. On the old grounds, however, they found a lower biomass and a more diverse fauna. *Hiatella* was much less common and shared dominance with sponges, *Halichondria* sp., anemones, *Tealia felina*, and urchins, *Strongylocentrotus droebachiensis*. The latter two species, it was suggested, were key species in the old ground community, the urchins grazing the sessile organisms, including *Hiatella*, and the anemones preying on the urchins. Both anemones and urchins were infrequent on the new grounds. Other examples of the important effects of urchins on their environments occur in later sections (e.g. § 5.5, 6.5).

Dramatic intermittent effects can be produced by large popu-

Fig. 3.1. Stages in the destruction of a mussel bed (*Mytilus edulis*) by starfish (*Asterias rubens*) at 15–20 m off Portland Bill, English Channel. (*a*) part of the mussel bed which has not yet been invaded by starfish; (*b*) active predation; (*c*) all mussels have been removed leaving bare rock with scattered, dense aggregations of starfish. All photographs were taken on the same dive. Starfish were about 20 cm across.

lations of starfish. Sublittoral mussel beds are particularly unstable in areas where mussel-eating starfish occur (Sloan, 1981). The common starfish *Asterias rubens* may aggregate on mussel beds, or beds of other bivalves, in densities of up to 100 m^{-2}. The aggregations move slowly over the beds leaving bare substrate and empty shells behind. At a site in the English Channel I have seen at least two cycles in ten years of mussel bed development and destruction by starfish (Fig. 3.1). This type of interaction has led to the hypothesis that predation may increase the local diversity of a community. In the case of mussels and starfish it works because mussels go in for mass settlement and fast growth and are thus very successful spatial competitors (see § 3.3 below). They can rapidly take over a habitat and exclude other species. Removal of mussels by starfish is almost bound to lead to the development of a more diverse fauna. A similar argument can be used to account for the relatively high diversity on the old larval grounds, noted above; here the important factor is the prevention by the urchins of the development of a dense exclusive population of *Hiatella*. In some environments, however, grazing by urchins may lead to the eventual colonization of the substrate by a few distasteful species, thus reducing diversity. This was observed on pier pilings at

Beaufort, North Carolina, which were being studied for comparison with communities developing on nearby experimental panels (Karlson, 1978). The panels bore diverse, temporally unstable communities whereas the piles were dominated by a sponge and a hydroid, both rare on the panels. The explanation appeared to be that urchins, present on the piles but absent on the panels, grazed most of the sessile species but disliked the sponge and the hydroid; thus these two had, with time, spread to cover most of the surface of the piles. Another example of predator exclusion was observed on the legs of an oil rig off California. Here a thriving mussel population living on the legs was apparently protected from starfish predation by a zone of anemones growing around the base of each leg and deterring the starfish from climbing up.

In general, however, the effects of predators are not easy to deduce. Small predators such as nudibranchs and pycnogonids, which feed on coelenterates, bryozoans and sponges, are very common but hard to see and very hard to exclude experimentally so as to determine the effects of their predation. Another predatory effect that is easy to overlook and difficult to estimate is that of the various suspension and deposit feeders on settling larvae. Anemones present a further problem; they are often easy to see and count but natural feeding is so intermittent that information is slow to collect.

Some interesting observations on natural predation by anemones were made by divers in the Antarctic (Dayton *et al.*, 1970). Two species, *Isotealia antarctica* and *Urticinopsis antarctica*, living on rocky substrates between 15 and 30 m deep (Fig. 1.2), were seen to catch large medusae when these strayed close to the bottom. *U. antarctica*, however, mostly fed on the urchin *Sterechinus neumayeri*. These urchins protect themselves with a covering of bivalve shells and other debris; when one is seized by an anemone, it simply releases its covering and escapes, leaving the anemone clutching the debris. Sometimes the protective covering includes a mat of living hydroids, these help by stinging the anemone, which retracts its tentacles. In an experiment, 80% of covered urchins escaped but all uncovered urchins were eaten.

Little is known about the effects of larger predators, such as fish, crabs and lobsters, on the nature of sessile animal communities.

These predators generally have a wide diet but most do not appear to feed directly on the sessile animals (except in the case of large crustaceans feeding on bivalve molluscs, serpulid worms and barnacles); rather they feed on other, smaller, mobile animals. Thus a fish might feed on smaller fish which might feed on scale worms or amphipods. In some cases, however, large predators control the populations of important grazers such as urchins, and so have far-reaching effects on the community (see § 5.5).

3.3   Competitive interactions

Once the larva of a sessile animal has settled on the substrate, its future growth and development depend on the vagaries of the physical environment, its chances of being eaten or parasitized, and its capacity to avoid being overgrown and smothered by its sessile neighbours (Fig. 3.2). This last possibility has been studied in animals growing on a variety of natural and artificial surfaces ranging from the undersides of plate-like corals to the surfaces of plastic panels. The pattern that is emerging is that overgrowth relationships exist and may be hierarchical, e.g. species A overgrows species B which overgrows species C ..., or, more commonly, may form a network, e.g. A overgrows B overgrows C overgrows A... Frequently there are twenty or more species involved in any given habitat and overgrowth relationships therefore tend to be complex. However, even simple relationships cannot be used to predict the outcome of competition since factors such as differences in rates of overgrowth between competitors, the arrangement of species with respect to each other (Fig. 3.3), and the available area of the substrate (large rock face, small isolated stone) all exert effects. The area is important since the greater it is, the more species are likely to be present and the more likely it is that a given species will meet a superior competitor. The result, or 'climax' community, is a dynamic state in which no single species triumphs and into which the chance occurrences of physical disturbance or predation constantly inject a random pattern of spaces which are available for new colonization.

Little work has been done on climax communities to assess the extent of this dynamic stability. Panel studies usually follow succession for only a few years and it is not known how long it takes for a climax to become established. A few studies of marked sites assumed to be supporting climax communities have, however, been carried out. Gulliksen (1980) described one such study on a rocky substrate at 8–14 m deep in a Norwegian fiord. Marked areas were photographed at intervals over a period of about four years and the total of 1698 photographs were analysed to follow the growth and turnover of the sessile fauna, the identity of which was confirmed by physical sampling of nearby areas. Important organisms included polychaetes and amphipods living in mud tubes, serpulid polychaetes, solitary ascidians, sponges and bryozoans; predators included urchins and starfish. Clear seasonal trends (e.g. Fig. 3.4) were detected and were repeated from year to year in the first half of the study, but a major destabilizing organism, the solitary ascidian *Ciona intestinalis*, was also present and produced a dramatic change during the second half of the study by a sudden very heavy settlement from the plankton (Fig. 3.4). Heavy settlements of *Ciona* had been seen previously in the same area, turning the bottom into a 'green carpet' (Gulliksen, 1980); these ascidians grow fast, they crowd out other epifaunal elements and, by

Fig. 3.2. Scanning electron micrograph showing an overgrowth interaction between two sheet-like bryozoans on the underside of a boulder. *Membraniporella nitida* (right) is overgrowing *Microporella ciliata*. Photograph by J. Rubin.

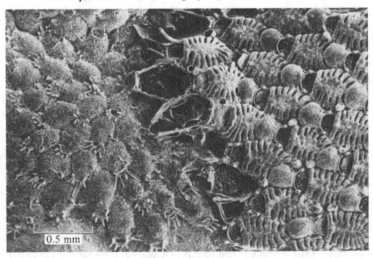

0.5 mm

Fig. 3.3. Diagrammatic representation of overgrowth dynamics on a spatially limited substrate in a simple network community where species A overgrows species B which overgrows C which overgrows A. Reading from left to right it may be seen that any one of the species may eventually dominate, depending on the initial arrangement on the substrate. After Buss & Jackson (1979).

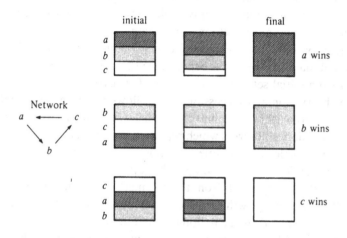

Fig 3.4. Hypothetical fluctuations of the most conspicuous groups in Borgenfjorden. An apparently stable, seasonally cyclic community is destabilized by a massive settlement of the ascidian *Ciona*. From Gulliksen (1980).

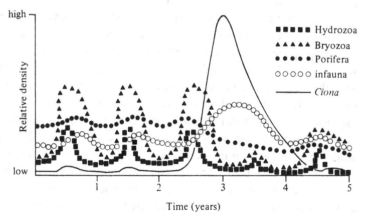

trapping sediment between individuals, improve the environment for infaunal species (Fig. 3.4). Predation by starfish and death of old individuals (*Ciona* lives 1–2 years) clears the area fairly rapidly and the community can then return to its original seasonally cyclic state. Sporadic mass settlement by organisms such as *Ciona* and *Mytilus* (see § 3.2 above) is evidently a very important factor in the ecology of some areas and only in places where such settlements do not occur can one expect long-term stability.

Jackson (1977) has suggested that in stable conditions colonial animals have a competitive edge over solitary forms during succession from initial settlement to climax. This hypothesis is based on the observation that succession often follows a pattern in which the primary colonizers are solitary – barnacles, serpulids, oysters – but the secondary colonizers which replace them and remain as dominants in stable communities – sponges, bryozoans, hydroids, colonial ascidians – are colonial. The supposed competitive advantage is thought to result from the colonial animals' indeterminate, multidirectional growth: a single colony can spread to cover a large area, death need only be regional, and directional growth gives the possibility of locating an environment more suitable than the one originally selected. Indeterminate growth also provides great potential for variation in form, and different forms may be viewed as serving different competitive strategies. Jackson (1979) defined three colony forms in which organisms are broadly attached to the substrate – runners, sheets and mounds – and three in which they extend into the water column – plates, vines and trees (Fig. 3.5). He suggested particular competitive advantages for each. Runners and vines, as their names suggest, rapidly spread through the surrounding space; they often occur as primary colonists along with solitary animals. Sheets exploit less space per gram of tissue but the area they occupy is entirely theirs, leaving no room for the settlement of alien larvae. Mounds, plates and trees are stages in the exploitation of the water column and escape from the competition on the substrate; however, their narrower connexion to the substrate makes them more vulnerable to physical disturbance or predation. All six forms are found commonly in bryozoans and sponges; hydroids mostly grow as runners, vines and trees; colonial ascidians usually form sheets but sometimes grow into mounds or plates. Individual species can sometimes adopt more than one of these forms: some bryozoans show runner-like growth in unfavourable environments and sheet-like growth under more favourable conditions. This illustrates the use of runners as exploratory devices for locating the most favourable local environment. Many hydroids show a combination of runners and trees; runners (stolons) creep over the substrate and produce upright hydranth-bearing trees at intervals. The stolons may eventually be overgrown by sheets of ascidians or sponge but the uprights remain and flourish.

The weapons used by spatial competitors to fight their border disputes vary from species to species and from group to group and are not yet well understood. Coelenterates use their nematocysts for offence and defence (see § 8.2 for corals). In sheet-like bryozoans the angle at which one colony meets another is important,

Fig. 3.5. Diagrammatic plan views and side views of six possible colonial growth forms. *A, a,* sheet; *B, b,* runners; *C, c,* mound; *D, d,* plate; *E, e,* vines; *F, f,* tree. Substrate sections are shown as solid bars. For further details see text.

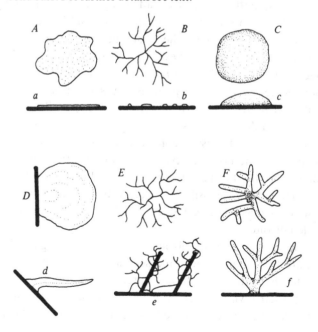

and structural differences such as colony thickness and lophophore height may play some part. Some sponges secrete toxic chemicals, while gelatinous colonial ascidians can avoid being overgrown by oozing round their competitors. Few of these techniques are available to solitary animals other than anemones. Most have inert shells over which colonies readily grow, and eventually the apertures through which they feed become occluded unless their growth can keep pace with that of the surrounding colonies. Serpulids faced with overgrowth of the aperture sometimes direct the growth of their tubes away from the substrate up into the water column. Clearly, however, solitary animals do survive, and may flourish. Some, e.g. large oysters, are long-lived and tolerant of overgrowth, even apparently encouraging it for the sake of camouflage; others depend on the creation of free space and a capacity for rapid primary colonization from the plankton. These are common in unstable habitats where free space is most frequently generated.

3.4   Commensal interactions

It is evident that space is at a premium in sessile communities. Consequently there is an advantage to be gained by any animal which can use another organism as a substrate. This is not necessarily a good thing for the 'host' but most of the larger sessile animals support, whether they like it or not, a variety of smaller hangers-on collectively referred to as epizoites. Epizoites are either sessile – hydroids, bryozoans, mussels, barnacles, etc. – or mobile – crinoids, brittle stars, shrimps, fish, etc. They are commonest on foliaceous hosts such as large erect bryozoans or hydroids (Fig. 3.6). Sessile epizoites are usually restricted to the inert parts of the host: the bases of hydroid, bryozoan and coral colonies and the shells of sessile bivalve molluscs are often heavily overgrown. Large sessile animals which lack exposed inert surfaces, such as sponges and octocorals, are mostly colonized by mobile epizoites.

In the case of sessile epizoites, the advantages gained by the habit are first, living space, and second, elevation which exposes them to the benefits of faster water flow without the need to develop their own support. This latter advantage also applies to mobile passive suspension feeders such as crinoids, which often move to elevated positions on the tops of sponges or gorgonians before spreading their filtration fans, and caprellid amphipods which cling to the upper branches of hydroids (Fig. 3.6), bryozoans

Fig. 3.6. Arrangement of epizoites on clumps of the host hydroid *Nemertesia*. On the inert holdfast there are deposit feeders (rissoid gastropods, the tube-dwelling amphipod *Corophium*) and active suspension feeders (sponges, sabellid worms, ascidians, the bryozoan *Scrupocellaria*). On the inert proximal hydrocauli are active suspension feeders (*Scrupocellaria*) and passive suspension feeders (the hydroid *Obelia*, the tube-dwelling amphipod *Erichthonius*). On the live distal hydrocauli are passive suspension feeders (the hydroid *Plumularia*, the amphipod *Caprella*). Details of this association were described by Hughes (1975).

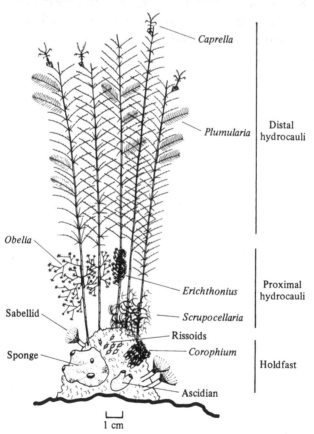

and algae while combing the current with their setose antennae. For many epizoites, however, the main advantage of the association may be protection from predation. Branched or labyrinthine host colonies provide good cover for soft-bodied organisms such as worms, and for small or juvenile crabs and fish. Bryozoan or hydroid thickets may be used as protected settlement sites by sessile animals. In the case of coelentrate hosts, the nematocysts no doubt help to deter some would-be epizoites but provide protection for those which live amongst them, such as the commensal shrimps, crabs and fish commonly associated with anemones and corals in tropical environments. These latter associations are often apparently obligate for the epizoite and some may be symbiotic (see § 8.7). Most epizoites, however, are facultative associates and may be found living on a range of different hosts as well as on otherwise uncolonized rock surfaces.

Part II: Kelp Forests

4

The plants

4.1　Introduction

Kelps are large brown algae belonging to the Laminariales. Most species are perennial and may live for ten years or more. They can be found near low spring tide level on most rocky shores outside the tropics but their main habitat is in the sublittoral where they form extensive forests. Consequently the bulk of our knowledge of kelp forest ecology comes from diving. Individual plants consist of a holdfast, one or more stipes or stalks and one or more blades; stipe and blades together may be referred to as a frond (Fig. 4.1). The holdfast is composed of rootlike haptera, the tips of

Fig. 4.1. Kelp morphology. (*a*) palmate, e.g. *Laminaria hyperborea, L. digitata*; (*b*) strap-like, e.g. *L. saccharina, L. longicruris*; (*c*) laterally branched, e.g. *Ecklonia* spp.

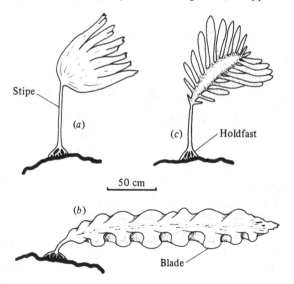

which adhere strongly to rock. Important forest-forming kelps include the giant kelp, *Macrocystis*, which may grow to over 50 m long, and several species belonging to the genera *Laminaria* and *Ecklonia*. Species of the latter two genera are usually no more than 2–4 m long but some species, e.g. *L. pallida* and *E. maxima*, may be 5–10 m long. Forests may be monospecific but more often two or more species occur together; the relative proportions of the different species vary with depth, wave action, etc. Kelps are mostly restricted to the cooler coastal waters of the world and do not generally stray into water that is warmer than 20°C in the summer. *Macrocystis* occurs around the southern hemisphere and off the west coast of North America. *Laminaria* is common in the North Atlantic and in the Pacific from Japan northwards; it is not an important kelp off the Pacific coast of North America. *Ecklonia* occurs around the southern hemisphere and off Japan.

In a large *Macrocystis* plant the conical holdfast is a metre or more in diameter and gives rise to more than 100 stipes. A fully grown stipe, although only about 1 cm thick, may be 50–70 m long. Blades, each about 1 m long, develop at intervals along the stipes; they become more frequent towards the tip of the stipe and tend to die off at the base. A single stipe may bear more than 200 blades. At the base of each blade there is a pneumatophore which provides

Fig. 4.2. A giant kelp (*Macrocystis*) forest.

buoyancy. The pneumatophores and long stipes of the giant kelp ensure that most of the blades are supported at or near the sea surface producing a substantial canopy to the forest. This canopy casts a gloom in which divers can swim freely amongst the columns of stipes (Fig. 4.2).

Forests of *Laminaria* and *Ecklonia* are usually much less roomy. The stipes of the major European species, *L. hyperborea*, are about a metre long and arise like the trunks of trees from the rock (Fig. 4.3). Each stipe supports a single palmate blade and these form a canopy which follows the contours of the rock but is 1–2 m above it. It is possible, as a diver, to creep along between the stipes and below the canopy rather as a large dinosaur might creep through a beech wood.

Four major contributions to kelp forest research have been made. Forests of *L. hyperborea* around British coasts have been extensively studied by Kain, and a series of papers has been published. This work culminated in a review of the genus *Laminaria* (Kain, 1979) in which other work on kelp, much of it on European species, was referred to. Two team efforts on production and energy flow in kelp forests are those of Mann's group on

Fig. 4.3. Under the canopy of a *Laminaria hyperborea* forest. The stipes are about 1 m high. 7 m, South Cornwall.

Laminaria longicruris in St Margaret's Bay, Nova Scotia (e.g. Mann, 1973), and Field's group on *L. pallida* and *Ecklonia maxima* off the Cape in South Africa (e.g. Field *et al.*, 1977; 1980). Both research programmes have led to the publication of a series of papers. A further team effort on the giant kelp beds of California arose out of the need to conserve the beds as an important natural resource. Kelp is valuable stuff, you can use it to make salad oil, tooth paste, ink, paint, animal food, human food, natural gas and many other things. Kelp beds also attract fish and lobsters and so are well worth conserving. Study on Californian *Macrocystis* during the 1960s resulted in the publication of numerous papers and three books, the final one of which (North, 1971) comprises a series of papers each on a separate aspect of the biology of the kelp bed ecosystem. A recent general review of kelp bed ecology is that of Mann (1982).

Most of what I have to say in the following two chapters is drawn from these sources, supplemented by more isolated research papers and by a general personal view gained by swimming about under the kelp canopy off the coasts of Britain. In this chapter I will concentrate on populations, growth and production of the plants, and the effects on plant populations of the major environmental variables of light, water movement and competition. In the following chapter, on the fauna, I will examine the environment provided for animals by kelp, describe the inhabitants, and the effects, if any, that they have upon the kelp.

4.2 Populations and biomass

Population density in kelp beds under favourable environmental conditions depends on the species and on the average age of the plants. In mature *Macrocystis* beds off California the average density is 0.5 plants m^{-2} with frond density varying from 2–15 m^{-2}. Settlement on bare rock in the absence of a mature canopy, however, can result in populations of up to 200 juveniles m^{-2}. Californian *Macrocystis* beds show cyclical changes in density from fairly sparse to fairly dense with a period of 4–5 years. *Macrocystis* beds off Argentina, however, have a shorter, more drastic cycle of 3–4 years from settlement to complete destruction by storms. These populations are composed of younger plants with a higher average density of 1–2 m^{-2} but with a similar frond density of 3–18 m^{-2}: younger plants have fewer fronds (Barrales & Lobban, 1975).

Small kelps can reach much higher densities than populations of giant kelp; e.g. 600 m^{-2}. Even in the moderate sized *Laminaria hyperborea*, 100 plants of all ages per square metre have been recorded. More typically in this species, however, there are 10–15 large canopy plants per square metre with up to about twice that number of smaller understorey plants. Populations of *L. hyperborea* are often distinctly bimodal with respect to stipe length and this is thought to be due to shading by the canopy inhibiting the growth of understorey plants. The surface area of the blades which form the canopy in a mature forest is often about five times the area of the substrate and measurements have shown that light intensity beneath the canopy is only 3–11% of that available just above it. Fig. 4.4 shows size frequency and age frequency histograms for some populations of *L. hyperborea* at different depths and the two shallower populations show a clearly bimodal size distribution which is greater than can be explained by the slightly bimodal age distribution. Age distribution is not typically bimodal – as the age

Fig. 4.4. The age and size structures of *Laminaria hyperborea* populations from the Outer Hebrides with relation to depth. For further explanation see text. From Kain (1977).

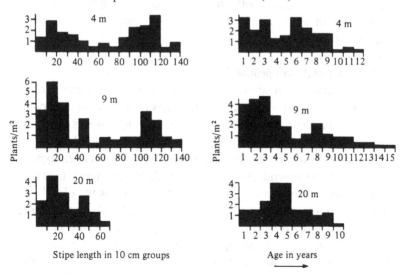

Stipe length in 10 cm groups   Age in years

frequency of the deeper population shows – indeed, as might be expected from the continuous chance of mortality, there is often a bias in favour of young plants. Shading breaks down in less dense forests and when canopy area is less than about twice the substrate area no relative inhibition of growth is found. Shading also occurs in *Macrocystis* forests in which, when the canopy is complete, recruitment of juveniles is nil. Recruitment only occurs when mortality produces gaps in the canopy through which light can penetrate to the sea bed. Similarly in *L. hyperborea* forests breaks in the canopy caused by the death of larger plants stimulate the growth of understorey plants which quickly grow up to fill the gaps. Frond length is not bimodally distributed in *Macrocystis* forests since all fronds reach the canopy.

Biomass measurements show less range than population density measurements. 3–22 kg fresh weight m^{-2} is found in *Macrocystis* beds off California and 6–7 kg m^{-2} off Argentina. *Laminaria* spp. generally show standing crops of 10–20 kg m^{-2}: *L. hyperborea* biomass in the Northeast Atlantic is often about 10 kg m^{-2} and *L. pallida* off the Cape of South Africa is about 14 kg m^{-2}. The mixed population dominated by *L. longicruris* off Nova Scotia, however, was found to have a standing crop of 20–29 kg m^{-2}. Right outside these ranges is a record of what must have been an impenetrable jungle of *Macrocystis* off the Kerguelan Archipelago in the Southern Indian Ocean. Here there were 20–290 fronds m^{-2} and a standing crop of 95–606 kg m^{-2}!

4.3 Growth and production

Kelps often show seasonal growth, and in the North Atlantic the season of rapid growth starts in the winter and continues through spring into early summer. *L. hyperborea* is the most extreme case, it grows rapidly from February to June but from June to January shows practically no growth. Rapid growth in *L. hyperborea* leads to the production of a new blade from a meristem at the base of the blade. Growth of the new blade pushes the old blade away from the top of the stipe. By March/April considerable growth has already occurred and the appearance of the plant is shown in Fig. 4.5. with the top of the stipe sporting a gleaming new blade, from which hangs the dishevelled remnant of last year's blade. The old blade eventually drops off or is torn away by wave action and adds its contribution to the detritus food chain. *L. longicruris* off Nova Scotia also grows most rapidly during late winter and spring but in this species a reduced rate of growth continues through the rest of the year. Growth has been measured in *L. longicruris* by punching holes in the blades of marked plants and following the movement of these holes along the blade (Fig. 4.6). It has been found that continuous growth of new blade from the base is offset by continuous erosion of old blade at the tip. In spring and summer, growth exceeds erosion and the blades may become very long, in the autumn erosion increases and some reduction in blade length may occur despite the fact that the blade is still growing. It has been estimated that *L. longicruris* plants renew their entire blades between one and five times per year. *L. pallida* off South Africa has similar 'conveyor belt' blades which grow continuously at the base and erode at the tips. These blades, aided no doubt by the well-fertilized upwelling waters off the Cape and the abundance of light, replace themselves six times per year.

Fig. 4.5. The appearance of a *Laminaria hyperborea* frond in early spring following the growth of the new blade. Last year's blade, fouled with epiphytes, has not yet been detatched and the sheet-like bryozoan *Membranipora* is 'escaping' by orientated growth from the old blade onto the distal end of the new.

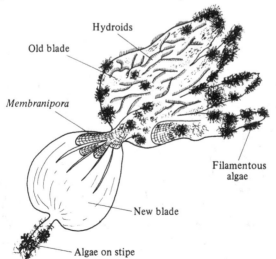

In higher latitudes the early start of rapid growth, at a time of year when there is little light available for photosynthesis, poses the problem of how the plants fuel their growth. Experiments have shown that *L. hyperborea* can grow a new blade entirely in the dark, but even in the light can only produce a half-sized new blade if the old blade is amputated before the start of growth. Thus most of the energy required for growth is translocated from existing tissues – old blade and stipe – where it was presumably stored from the last summer's photosynthesis. Another factor supporting rapid growth in high latitudes during winter and spring is the ready availability, at this season, of inorganic nutrients, which are often depleted in surface waters later in the year. Finally, it is clearly advantageous to produce a large photosynthetic area early in the year so as to be ready to take advantage of higher light levels as soon as they become available.

Rates of blade growth attained during fast growth in various *Laminaria* species vary from 66 mm week^{-1} in *L. hyperborea* in the Irish Sea, through 121 mm week^{-1} in *L. saccharina* off Scotland, to 250 mm week^{-1} in *L. augustata* var. *longissima* off Japan. Stipes grow at the same time as blades, but not being subject to erosion

Fig. 4.6. Diagram to show how growth can be measured in straplike kelps. From Mann *et al.* (1979).

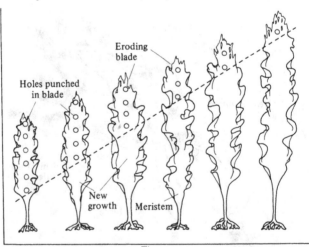

they grow less. The maximum observed growth rate of *L. hyperborea* stipe is 10–13 mm week^{-1}. A final addition by growth is to the holdfast. In *L. hyperborea* new haptera sprout each year from around the base of the stipe (Fig 4.7). Their growth takes them

Fig. 4.7. Holdfasts of *Laminaria hyperborea* in July showing growth of new (unencrusted) haptera. Encrustations visible on old haptera include white colonial ascidians, pale grey sheet-like bryozoans and dark grey sponges, much silt is also present. Red algae are plentiful on the bases of the stipes. 7 m, South Devon.

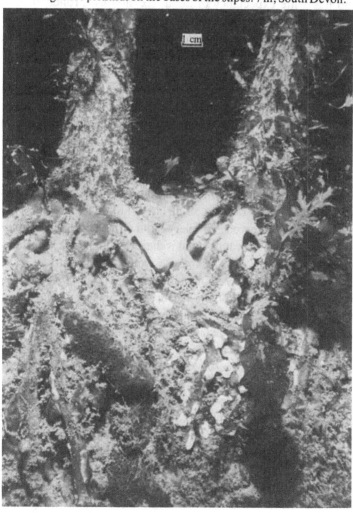

curving over the old haptera to meet the rock at the outer edge of the holdfast. This annual growth of haptera can be used to age plants of *L. hyperborea*: a longitudinal section of the holdfast reveals annual layers or whorls of haptera, often with corresponding growth lines showing annual additions to the thickness of the stipe (Kain, 1963).

Growth in *Macrocystis* is rather different from that in *Laminaria* since in giant kelps entire fronds are renewable: in California each frond has a life expectancy of about six months. Fronds arise in pairs from just above the holdfast and grow rapidly to reach the water surface where the pneumatophores keep the blades floating in well-lighted water. In California growth is continuous at a steady high rate throughout the year. A six-month-old plant living in 10–15 m deep water has usually had time to grow two fully developed fronds with second and third pairs partially grown. After a year's growth there are generally about eight fully developed fronds with one or two pairs already becoming senile and deteriorating. A two-year-old plant has about 25 fronds and a conical holdfast 20–30 cm in diameter. The most rapid frond growth occurs when the fronds are 10–20 m long – about half of their mature length. 30 cm day^{-1} is an average maximum rate but 50 cm day^{-1} (10–15 feet week^{-1}) has commonly been observed. This very rapid growth often takes place in the shade below the canopy and is therefore fuelled by translocation of energy from existing mature fronds on the surface. The growth rate of fronds of young plants less than one year old is 10% slower than that of older plants because the young plants have relatively fewer mature fronds to supply this energy. California, like the Cape, is probably an ideal environment for kelp growth with its cool, nutrient-rich water well-lighted all the year round. Further north off Vancouver, *Macrocystis* frond growth is seasonal, in the spring, and slower, about 40 mm day^{-1}.

The productivity of kelp beds is remarkably high and the highest figures are similar to the highest terrestrial productivities. At the lower end of the scale *L. hyperborea* production, measured by cropping the new blade at the end of the seasonal growth spurt, is between 0.4 and 0.7 kg C m^{-2} yr^{-1}. *L. pallida* off South Africa achieves about 1.3 kg C m^{-2} yr^{-1}. while the mixed population in St Margaret's Bay, Nova Scotia, dominated by *L. longicruris* produces 1.75 kg C m^{-2} yr^{-1}. These productivities were estimated from blade growth measured by punched holes. *Macrocystis* production off California is in the same high range: 0.4–0.8 kg C m^{-2} yr^{-1}. Kelp bed production is 5–10 times higher than nearshore phytoplankton production and indicates the probable importance of these seaweeds in supporting nearshore food chains.

4.4   The effects of depth

The main effect of depth is on the amount of light reaching the sea bed. Light decreases logarithmically with increasing depth and the lower limit for kelp growth has been found to be the depth at which, averaged over the year, about 1% of the surface light reaches the bottom. This depth varies with the clarity of the water; the deepest kelp is in the clear waters of the Mediterranean where *Laminaria ochroleuca* forests occur at 80 m in the Straits of Messina and *L. rodriguezii* forests at 90–110 m off Corsica. Kelp does not grow more shallowly in the Mediterranean because the surface water is often too warm. In the North Atlantic the water may be clear enough to support kelp growth down to 30 m below low spring tide level, but usually the lower limit is less than this and in the murky waters of harbours and estuaries it may be as little as 2–3 m below the low tide mark.

Before the lower limit is reached, the declining light has an effect on the rate of growth of kelp plants. This effect can be seen in Fig. 4.4 which shows that the deepest population of *L. hyperborea* has a lower population density and is composed of smaller plants than the two shallower populations. The age distribution shows that this is not a result of the plants being young: there are fewer of both young and very old plants in this deep population compared to the two shallower ones. Where *Laminaria* extends to 20–30 m deep the standing crop is usually a tenth or less of its value at 5–10 m (Fig. 4.8) and production is also correspondingly less. A further change in morphology with depth is that *Laminaria* blades tend to be thinner in deeper water.

Giant kelps have similar depth limits to those of the smaller kelps since, although mature fronds reach the surface, the young plants must start their growth in the light intensity prevailing on the sea bed. The reduced growth rate of young giant kelp has already been mentioned, but once the plant has several fronds at the surface the growth rate of subsequent fronds is independent of depth. The final length to which the fronds grow, however, is slightly more than double the depth, so the greater the depth, the more growth takes place. The deepest giant kelp forests are in 35–40 m of water.

Apart from light intensity, the other depth related factors include temperature, nutrients and wave action. These three factors may affect the upper limits of kelp forests. Experiments on summer growth of *Laminaria saccharina* transplanted from low spring tide level to depths down to 20 m showed that growth was maximum at 11.5 mm day^{-1} at 9 m, was much reduced at 17 m, but was only 5 mm day^{-1} at 0.1 and 3 m. This reduction in growth at high light levels was attributed to the shallower plants spending more time in the relatively warm, nutrient-poor surface water layer. The effects of wave action are discussed below.

4.5 The effects of water movement

Like light, wave action decreases logarithmically with depth. The effects of wave action are therefore most intense in

Fig. 4.8. Standing crops of populations of *Laminaria hyperborea* from the Outer Hebrides with relation to depth (see also Fig. 4.4). After Kain (1977).

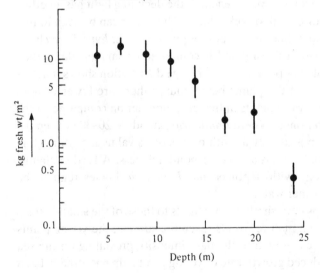

shallow water where fierce water movement can cause mortality by breaking stipes, tearing blades and wrenching holdfasts from the rock. Some kelp species are better able than others to tolerate these stresses. *L. digitata* has a more flexible stipe than *L. hyperborea* and a holdfast that spreads its attachment over a larger area of rock. Since it bends with the waves it experiences less drag than the stiff, upright *L. hyperborea*. On exposed coasts therefore, *L. hyperborea* may be replaced by *L. digitata* in the shallows just below low tide level. In addition to being flexible, *L. digitata* stipes need to be fairly strong in tension to resist the drag of the waves on the trailing blades; the tensile breaking stress of *L. digitata* stipe, however, is only about 5 MN m^{-2}; rather low for a structural biomaterial (Wainwright *et al.*, 1976). Giant kelps with their buoyant fronds need to cope with even greater tensile forces since as ocean swells pass they lift the floating canopy and tug on the stipes. Contrary to expectations Koehl & Wainwright (1977) found that the tensile breaking stress of the stipes of the giant kelp *Nereocystis* (3.6 MN m^{-2}) was no higher than that of *L. digitata* and the critical adaptation turned out to be extensibility rather than strength. Before it could break, the stipe had to be stretched by 30–45% of its resting length, and each tug of the swell hardly ever lasts long enough to achieve this degree of strain in the long (20 m or more) stipes. Thus the breaking stress, although low, is hardly ever reached. During storms, however, stipes of giant kelps may break (usually at flaws caused by an urchin bite or other abrasion) and holdfasts may be wrenched free. Detached plants may then become entangled with their neighbours, increasing the drag on individual holdfasts, and storms can thus rip up entire beds. This happens regularly off the rough coasts of Argentina, but less often off California. Mortality caused by wave action tends to restrict the longevity of kelps in shallow, exposed sites and populations in these environments are usually composed of relatively young plants.

Since wave-drag on kelp blades is a major cause of mortality in storms, an adaptation of growth form to ambient wave exposure might be expected. Gerard & Mann (1979) examined exposed and sheltered populations of *L. longicruris* and found characteristic differences in growth form (Fig. 4.9). Transplant experiments showed first, that the sheltered growth form could not survive in the exposed habitat (the broad thin blades were torn to bits), and second, that the growth form was controlled by the environment since exposed plants, transferred to the sheltered habitat, began to grow sheltered-type blades. Gerard & Mann also found that growth rates were generally lower in the exposed plants and suggested that this was because of the reduction in photosynthetic and nutrient-absorbing surface consequent upon the thickened narrow blade-form.

The exposed population examined by Gerard & Mann (1979) was subject to severe wave action but moderate water movement, either as waves or currents, appears to be beneficial to kelp growth and production. Water movement prevents silt settling on the blades where it might screen off some of the light, and prevents siltation of the rocks on which the populations grow. Some siltation of holdfasts can occur without damage to mature plants but the tiny gametophyte generation may be killed by as little as a 0.5 mm layer of sediment. Further, waves and currents increase the inorganic nutrient supply to the plants and, in general, production seems to be slightly enhanced in moderately exposed kelp beds.

4.6 Plant competition

Some indication of plant competition has already been given with the description of the competitive edge over *L. hyperborea* which wave action gives to *L. digitata* at and just below low tide

Fig. 4.9. *Laminaria longicruris* from exposed and sheltered sites showing typical morphology and posture. Sheltered plants have broad thin blades whereas exposed plants have narrow thicker blades with a corrugated marginal frill. From Gerard & Mann (1979).

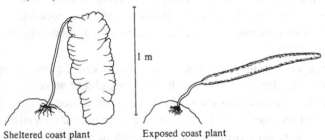

Sheltered coast plant   Exposed coast plant

1 m

level in the Northeast Atlantic. Similar zonation, probably also controlled by wave action, is found off the Cape where *Ecklonia maxima* occurs in shallow water and *Laminaria pallida* offshore; and again off California where giant kelps are replaced by other smaller species in water shallower than 5–10 m. The influence of wave action on Californian zonation can be seen in sheltered areas where giant kelps extend up into the low intertidal. In the Northeast Atlantic on fairly sheltered coasts *L. hyperborea* may similarly appear in the low intertidal, but this species grows where there is moderate water movement and is replaced in very sheltered environments by the strap-like *L. saccharina.*

What gives the dominant species the competitive edge in deeper water? The answer appears often to be size, shape and longevity which together lead to shading. Mature *Laminaria hyperborea*, with its stiff, upright stipes, shades the lower growing *L. digitata*; giant kelps shade all smaller species. The influence of longevity can be seen when plant succession on cleared areas is studied. Cleared areas of rock in an *L. hyperborea* forest off the Isle of Man were quickly colonized by a variety of algae including *L. hyperborea*. After one year *L. hyperborea*, *L. digitata* and *L. saccharina* had all reached a similar size but were all overshadowed by the faster growing annual species *Saccorhiza polyschides*. After two years the *L. hyperborea* plants had reached a greater height than the other *Laminaria* spp. and were producing sufficient shade to reduce the recruitment of *Saccorhiza*. After four years the cleared areas were indistinguishable from virgin *L. hyperborea* forest except in respect of the ages of the canopy plants.

In sheltered areas such as St Margarets Bay, Nova Scotia, a different situation obtains. Here the palmate *L. digitata* is replaced in deeper water by the strap-like *L. longicruris*, a species unlikely to succeed by shading its competitors. Both *L. longicruris* and *L. saccharina* appear to do well in sheltered environments but the adaptations which help them to do so are not clear. The strap-like shape is probably not the answer since *Alaria esculenta*, a small North Atlantic kelp characteristic of extreme exposure, is also strap-like. Possibly *L. longicruris* is better adapted to low light intensities than *L. digitata*. This type of adaptation must be important at the lower limits of kelp forests where red algae, common as understorey plants within the forests, replace the kelps as the dominant flora on open rock surfaces. These algae include foliaceous forms and encrusting corallines; being red they reflect practically no light (red light is rapidly absorbed by water and is virtually absent below about 10 m) and so are able to use a high proportion of available wavelengths.

Another influence on the composition of the algal community is grazing pressure. The chief grazers are sea urchins (see below, § 5.5) which, if they have preferences amongst the available food plants, could affect the relative proportions of different species. The subject of urchin food preferences has been reviewed by Lawrence (1975). From the point of view of the plant, the competition would be won by the most distasteful species (cf. § 3.2). Apparent instances of this effect have been observed in competition between species of *Laminaria* and *Agarum* in which the latter replaces the former in the presence of grazing urchins. The effect depends, however, not just on distastefulness, but also on the interplay between plant growth and urchin grazing rate, and on wave action. The deeper the water, with its consequence of slower growth, and the greater the urchin density, the greater the disadvantage of the grazed species. Conversely, the shallower the water, with its consequence of greater wave action, the less easy it becomes for the urchins to hold and graze the writhing fronds. The wider effects of grazing, especially by urchins, are discussed later.

5

The fauna

5.1 Introduction

The animals that inhabit kelp forests include those, usually small, forms which live on the surface of the kelp itself or amongst the haptera of the holdfast, and those, usually larger, species which live on the rock or in the water amongst the fronds (Fig. 5.1). Both groups contain sessile and mobile members. Amongst the small close associates are forms such as hydroids encrusting the blades and stipes, and crabs and brittle stars living in the holdfasts between the haptera. The larger associates include sponges living attached to rock, and fish, lobsters and urchins moving about within the forest. Most of these animals are not restricted to the kelp forest environment but also occur on rocky substrates in deeper water; often, however, they are commoner in the kelp than elsewhere and a few species only occur amongst kelp. Restriction to kelp, where it occurs, is usually a result of feeding preferences; rather few species, however, graze the kelp directly and one of the most important groups of kelp grazers – the sea urchins – also grazes detritus and sessile organisms other than kelp, and thus can exist outside the forest.

The two commonest reasons for inhabiting kelp forests are, first, the use of the forest as a shelter from predators, water movement, etc., and second, the use of the plant surfaces for attachment. Shelter occurs at two levels – the large scale of the whole plant and the smaller scale of the holdfast. Algal surfaces provide suitable attachment sites for a wide range of species, some of which are limited to algae. Of the plant surfaces, the most easily colonized are the holdfasts, and some kelps, such as *L. hyperborea*, have rough stipes which encourage settlement of sessile organisms. Blades are less easily colonized and support a more specialized assemblage.

The fauna   62

In this chapter, rather than deal with the fauna according to way
of life – encrusters, grazers, shelterers – I will describe them according to community – holdfast, frond, forest. The exception is the
important grazing group of the urchins, which deserves special
treatment. Lastly it is worth considering the implications of the
high primary productivity of the forests and the fact that little of it is
directly grazed by herbivores. Energy flow studies by Mann's
group in Nova Scotia and Field's group off the Cape have high-

Fig. 5.1. Pictorial representation of the fauna of a *Laminaria
hyperborea* forest. Kelp blades support sheet-like bryozoans (1),
hydroids (2), and blue-ray limpets (3); stipes are colonized by red
algae (4), which are grazed by urchins (5). Sessile fauna on the
rocks include sponges (6), hydroids (7), anemones (8), fleshy
octocorals (9), branched bryozoans (10), mussels (11) and
ascidians (12). Mobile fauna include lobsters (13), crabs (14),
starfish (15), holothurians (16), conger eels (17), wrasse (18),
gobies (19) and gadoid fish (20) swimming above the canopy.

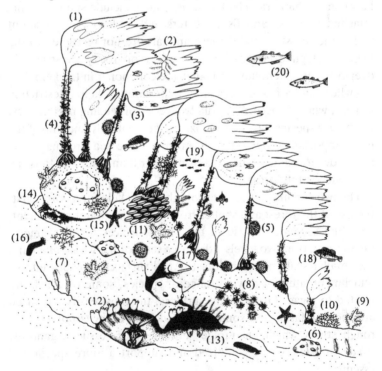

lighted the importance of kelp detritus and filter feeders in transfer-
ring energy to higher trophic levels.

5.2   Holdfast fauna

Darwin (1845) writing of *Macrocystis* off South America,
observed: 'On shaking the great entangled roots, a pile of small
fish, shells, cuttle-fish, crabs of all orders, sea-eggs, starfish,
beautiful Holothuriae, Planariae and crawling nereidous animals
of a multitude of forms all fall out together'. Of the common, easily
visible animals Darwin merely missed the amphipods and isopods
and, slightly less important, the brittle stars; judging from Californian *Macrocystis* he was over-enthusiastic about the cuttle-fish. In
Californian *Macrocystis* holdfasts amphipods, isopods and polychaete worms are the dominant groups and, in a single holdfast,
can be expected to account of 40–90% of all individuals. North
Atlantic *Laminaria* holdfasts contain a similarly diverse assemblage, with polychaetes and amphipods again being important. A
feature of the holdfast environment is the sediment that accumulates between the haptera. This provides a substrate in which
worms, and crustaceans such as tanaids, can dig burrows; other crustaceans such as tubicolous amphipods construct their tubes from
sediment and deposit feeders can use the organic fraction of the
sediment as food. Larger animals, however, such as brittle stars
and crabs, inhabit sediment-free spaces between the haptera and
may be excluded from heavily sedimented holdfasts.

A large *Macrocystis* holdfast may have a total volume (including
haptera) of 20–120 l and is composed of dead, dark brown haptera
in the centre, an intermediate layer of brown 'woody' haptera, and
an outer layer of young yellowish living haptera. Ghelardi (1971)
cut samples underwater from Californian holdfasts and found that
the inner and outer layers contained different assemblages. In 100
ml samples of the outer living layer an average of 48 individuals
belonging to 15 species was found, whereas in 100 ml samples from
the dead inner layer there were 79 individuals belonging to 35
species. Young, entirely living holdfasts were more populous (136
individuals/100 ml) but less diverse (22 species/100 ml) than the
inner regions of large holdfasts. In all cases the commonest groups
were amphipods, isopods and polychaetes.

Jones (1971), in a study of *L. hyperborea* holdfasts in the North Sea, measured the available space for colonization within a holdfast (between the haptera) and found that this 'ecospace' increased exponentially with holdfast age. Jones collected holdfasts of different ages and, by arranging them in a time series, showed that as the holdfast aged so the community became more diverse in a manner similar to succession (Fig. 5.2). One-year-old holdfasts are almost all surface area with very little ecospace, and were colonized by two species of encrusting bryozoans. Seven-year-old holdfasts, by contrast, have an ecospace of about 300 ml and supported an average of 80 individuals belonging to 19 species. Taking the holdfast community as a whole, a total of 53 species was found, of which 44 were recruited into the habitat in a distinct series and almost without exception remained in the community,

Fig. 5.2. The fauna of *Laminaria hyperborea* holdfasts from unpolluted (1) and polluted (2) North Sea populations. (*a*) number of species of associated animals, and (*b*) number of individuals of associated animals, with relation to holdfast age and habitable volume within the holdfast (ecospace). Each point in the mean of ten replicates ± standard error. From Jones (1971).

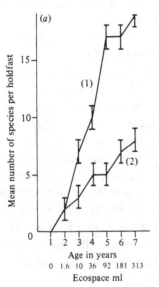
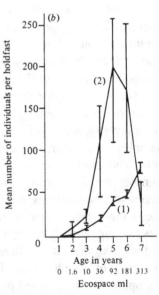

gradually increasing in numbers for the duration of habitat development. Important groups in order of decreasing abundance included brittle stars, both free-living and tubicolous amphipods, polychaetes, starfish, saddle-oysters, encrusting bryozoans, crabs, gastropod molluscs and urchins. As may be seen from Fig. 5.3, nearly half the individuals were suspension feeders.

Jones (1971) compared this 'normal' holdfast community, sampled on an unpolluted part of the coast, with the community in polluted kelp forests close to major urban developments. Polluted holdfasts provided the same amount of ecospace as unpolluted holdfasts, but contained a much less diverse fauna: an average of 44 individuals belonging to only eight species in seven-year-old holdfasts. Numbers of individuals in younger holdfasts, however, usually exceeded the numbers found in comparable unpolluted holdfasts (Fig. 5.2). The drop in numbers of individuals in seven-year-old polluted holdfasts may have been due to siltation of these much more capacious habitats: siltation would reduce water circulation in the interior. As a whole, the polluted holdfast community contained 44 species of which only 27 became progressively and permanently included as the holdfasts aged. The most abundant species were mussels and suspension-feeding polychaetes, and the dominance by suspension feeders was greater in the polluted environment (Fig. 5.3).

Fig. 5.3. Pie diagrams showing the percentage composition in terms of feeding category, of *Laminaria hyperborea* holdfast fauna from an unpolluted area (1), and a polluted area (2) in the North Sea. SF, suspension feeders; O, omnivores; C, carnivores; H, herbivores; DF, deposit feeders. From Jones (1971).

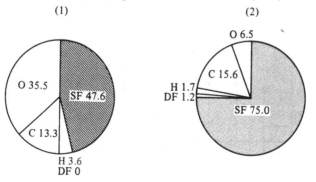

Although the bulk of the animals inhabiting holdfasts are suspension feeders, there are occasional herbivores feeding on the haptera and these are important because they weaken the attachment to the rock and so reduce the longevity of the plants. One of the commonest isopods in the holdfasts of *Macrocystis* is the kelp gribble *Phycolimnoria* which grows up to 6 mm long and chews galleries in the haptera. In California it is present in almost all holdfasts at a density of 5–20/100 ml holdfast sample. Gribbles are commoner in the outer live haptera than in the inner dead regions, but most of the dead haptera are hollowed out by earlier gribble grazing, indeed the gribbles are primarily responsible for the death of haptera. *Phycolimnoria* infestation combined with wave action is a major cause of plant mortality in the giant kelp beds.

In *L. hyperborea* a common grazer is the blue ray limpet *Patina*. This small mollusc, 5–10 mm long, mostly grazes on the blade but occasionally migrates down the stipe and into the holdfast. Here, at the base of the stipe, it hollows out cavities which appear to weaken the attachment of the stipe to the haptera. Evidence that this damage reduces the longevity of the plants is seen in the age-specific incidence of cavities which rises to a peak in middle-aged plants but declines amongst older plants (Fig. 5.4). The decline is interpreted as being due to increased mortality amongst damaged

Fig. 5.4. Percentage infestation by the blue-ray limpet *Patina pellucida* in different age classes of *Laminaria hyperborea* from moderately infested populations. Infested plants have one or more cavities in the holdfast. Modified from Kain (1963).

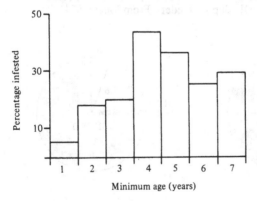

Minimum age (years)

plants. The extent of *Patina* damage varies from place to place, in some populations 50% of holdfasts contain one or more cavities, but 10–20% is more usual. Off Scandinavia *Patina* grazes blades but apparently lacks the habit of migrating into holdfasts.

5.3   Stipe fauna

Unlike the holdfast, the stipe does not provide extra shelter and so is colonized merely as a surface and mostly by encrusters such as bryozoans, sponges, colonial ascidians, hydroids, barnacles, etc. The texture of the stipe surface appears to influence its suitability for these encrusters since smooth stipes, such as those of *L. digitata*, are usually less well colonized than rough stipes such as those of *L. hyperborea*. Many sessile animals found on stipes are also found on surrounding rocks, but the stipe habitat is a favourable one since it is vertical and elevated from the substrate and so both avoids sedimentation and is exposed to water currents. Further, it offers some protection from benthic predators which may be reluctant to climb up (but see below, § 5.5). Smaller predators, however, such as nudibranchs and pycnogonids, may be found creeping about amongst their prey on the stipes.

5.4   Blade fauna

The blade is an unstable habitat compared with the holdfast and stipe. Not only is it constantly undulating in water currents, it is also a temporary surface, continually eroding and being replaced by growth. Despite this instability, kelp blades are colonized by a variety of animals from tiny crustaceans and foraminiferans through grazing molluscs and amphipods, to encrusting bryozoans and hydroids. The latter are important in providing extra structure in the habitat and in trapping detritus, thus encrusted blades support a richer fauna than clean blades. Since the distal parts of a *Laminaria* blade are older than the proximal parts there is usually a gradation towards greater encrustation and richer fauna from proximal to distal along the blade.

A study of the tiny animals on *Macrocystis* blades in California revealed that the most important were harpacticoid copepods, ostrocods, nematodes and turbellarians. Total numbers varied from about 100 dm^{-2} on clean blades to several thousand dm^{-2} on encrusted blades. These tiny organisms were mostly deposit feeders and may have been feeding partly on detritus and partly on resident bacteria. Bacterial populations on *Laminaria* blades have been studied off South Africa and densities at the proximal (young) ends of blades varied from 10^3 cm^{-2} in the winter to 10^6 cm^{-2} in the summer. At the eroding distal ends density was about 10^7 cm^{-2} all year round. Erosion of the distal blades was estimated to contribute 2.6×10^9 bacterial cells m^{-2} day^{-1} to the surrounding water.

In addition to the microscopic fauna and flora some larger solitary animals may be present such as, on *Macrocystis* blades, scallops, serpulid polychaetes, stalked barnacles, the kelp crab *Pugettia,* the isopod *Idotea* and the kelp curler *Amphithoe* (this amphipod folds or curls a flap of blade over itself for shelter). The latter three species are common grazers of giant kelp blades and *Amphithoe* in particular, is capable of rapid population growth leading to widespread destruction of blades by grazing. Under

Fig. 5.5. The sheet-like bryozoan *Membranipora* growing on a *Laminaria hyperborea* blade. The colony is growing towards the left, towards the proximal and youngest parts of the blade. On the right the oldest parts of the colony are fouled by algae and silt. In the central parts the evenly spaced spots are exhalant areas for the feeding current of the colony.

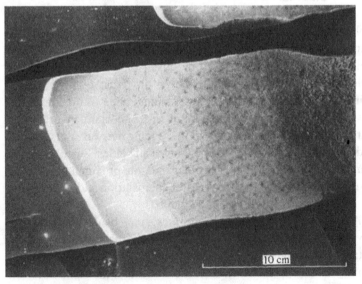

normal conditions, however, its populations, and those of other epiphytic animals, are kept in check by predatory fish. The little wrasse *Oxyjulus*, the senorita, is especially important in this context, as exclusion experiments have shown (Bernstein & Jung, 1979). In *Laminaria* forests there is much less animal diversity on the blades. The only common conspicuous animal, apart from encrusting colonies on *L. hyperborea* blades, is the blue ray limpet *Patina*. On *L. longicruris* the winkle *Littorina* may be common.

Perhaps the most important of the epiphytic animals on the blades are the colonial encrusters. Common species are the sheet-like bryozoan *Membranipora* (Fig. 5.5) and the hydroid *Obelia* (Fig. 5.6) which spreads by the growth of stolons. These encrusters may cover the entire surface of the blade and the weight of the encrustation may exceed the weight of the blade itself. It seems likely that such encrustations would cut down the rate of kelp photosynthesis but experiements on *Macrocystis* (Wing & Clendenning, 1971) have shown that at high light intensities, such as occur close to the sea surface, blades totally encrusted with

Fig. 5.6. The tip of a *Laminaria hyperborea* blade colonized by the hydroid *Obelia* which spreads over the surface of the blade by the growth of stolons.

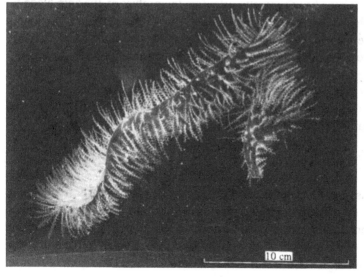

10 cm

Membranipora have the same photosynthetic capacity as do clean blades. (In fact they have rather more since *Membranipora* contains an associated green alga, the photosynthesis of which proceeds faster than the respiration of the bryozoan.) This is because close to the sea surface the photosynthetic process is saturated with light and the shading caused by the encrustation is insufficient to reduce the light below saturation. Experiments at lower light intensities showed a reduction in photosynthesis of encrusted blades. Encrustation probably has a negligible effect on the production of *Macrocystis* since mature encrusted blades mostly occur near the surface. In *Laminaria*, however, which may be similarly encrusted and in which mature blades grow close to the sea bed, encrustation may reduce total production and may limit the extension of forests into deeper water.

If encrustation is disadvantageous, as suggested above, one might expect adaptations by the kelp to discourage its occurrence and it is worth considering the rapid winter growth of *Laminaria* in this regard. Blade growth occurs at a time of year when encruster growth is relatively slow and when few encruster larvae are settling from the plankton. Thus the plant achieves a large area of clean blade. New blade produced later in the year is liable to be covered by encrusters almost as soon as it grows. Indeed many kelp encrusters have been shown to orientate themselves on the kelp blade to achieve maximum residence time. Larvae settle preferentially on the proximal, younger parts of the blade and established colonies show maximum growth towards the proximal regions. In March/April in the middle of the growth spurt of *L. hyperborea* this race for possession of the blade can be seen to good advantage with the encrusters desperately growing to avoid being shed with the old blade (Fig. 4.5).

5.5   Urchins

The importance of urchins as grazers on rocky substrates has already been stressed (§3.2) and kelp forests are, if anything, more sensitive to urchin grazing than are the deeper sessile animal communities since urchins often appear to prefer laminarian algae to other types of food (Lawrence, 1975). Urchin grazing has two effects on kelp forests. First, modest densities can greatly reduce kelp recruitment by grazing the rock surface between established plants and removing newly settled algae. Experiments in which urchins, *Echinus esculentus*, were removed and subsequently excluded from certain areas of rock off the Isle of Man showed greatly enhanced recruitment of *L. hyperborea*. The second effect occurs when urchins attack mature plants; if urchins are in sufficient numbers they may destroy entire forests of kelp. There is no clear separation between these two effects which depend partly on urchin density, partly on kelp growth rate, and partly on the preferences of the urchins. My own observations on *Echinus* off Cornwall, for example, suggest that this urchin rarely eats mature *L. hyperborea*. Its grazing activities on the rock between mature plants give the forest a park-like appearance, with few sessile organisms growing on the substrate (Fig. 4.3). Rather more organisms grow on the mature stipes, but even here they are not safe since I have observed urchins climbing up the stipes and even venturing onto the blades a metre above the rock to graze the epiphytic growth. Organisms such as red algae and barnacles were found in the guts of these urchins, but no *Laminaria*, and there was no evidence of urchin-damage to mature *L. hyperborea*. Where *L. hyperborea* growth is slow, however, *Echinus* may prevent the development of a forest by grazing all recruits before they have grown sufficiently large to be immune from attack. Thus the lower limit of kelp growth may be set by a combination of low light and urchin grazing, and this has been observed off the Isle of Man where, at sites with five or more urchins per square metre, the lower limit of the kelp is unusually shallow.

Where urchins feed on mature plants there is a risk that, if urchin population density increases, entire forests may be destroyed. Between 1911 and 1960 a kelp bed off California decreased in area from 5.9 to 0.088 square miles and urchin grazing was regarded as having played an important part in the destruction. *Macrocystis* in California is grazed by three species of urchin, *Strongylocentrotus franciscanus*, *S. purpuratus* and *Lytechinus anamesus*. Leighton (1971) discussed the general topic of grazing, especially by urchins, on Californian *Macrocystis* and described his observations on the destruction of an isolated, two hectare stand of giant kelp. Destruction was completed in three months by a grazing front of *S.*

franciscanus and *S. purpuratus* which moved through the kelp at about 10 m per month. The grazing front was mostly *S. franciscanus* at the leading edge and *S. purpuratus* at the trailing edge with peak density of the mixed population rising to about 60 m^{-2} (about 2.5 kg m^{-2}). The passage of the front across an established quadrat, and the simultaneous effect on the kelp, is illustrated in Fig. 5.7. After the front had passed, 'nothing remained on the bottom but sea urchins, dead central portions of holdfasts and the rock-encrusting coralline alga, *Lithothamnion*. This pattern of kelp bed destruction has been repeatedly seen . . .' (Leighton, 1971).

Apart from frontal attack, another way in which giant kelps are killed by urchins is by resident juveniles in the holdfasts chewing cavities in the haptera and so weakening them. Up to 100 small urchins may be found in a large holdfast and their activities, like those of the kelp gribble and the blue ray limpet, increase the chances of storm damage.

Kelp bed destruction has also been observed off Nova Scotia where it was estimated by Mann's group that 70% of beds had disappeared in the space of six years. Here the dominant kelp is *Laminaria longicruris* and the urchin is *Strongylocentrotus droebachiensis*. Kelp bed edges receding under urchin frontal attack were observed (Breen & Mann, 1976). Urchin density at the kelp edge

Fig. 5.7. Sequential sampling of a single quadrat (8.4 m^2) in a *Macrocystis* bed before, during and following the passage of a dense grazing front of urchins, *Strongylocentrotus* spp. From Leighton (1971).

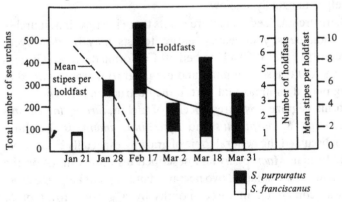

was about 200 m^{-2} and this part of the population was composed of relatively large individuals. Fig. 5.8 shows the relation found between urchin biomass and the rate of kelp bed recession; a critical point seems to be reached at 2–3 kg urchins m^{-2}, the same biomass that was observed in the Californian grazing front. Breen & Mann suggested that destructive grazing was only possible when there were sufficient urchins to climb onto the kelp blades and weigh down and immobilize the plants. Thus aggregation, and even the large size of the individuals, are important attributes of the grazing front. Further work in Nova Scotia by Bernstein *et al* (1981), however, showed that aggregation in these urchins is a form of anti-predatory behaviour rather than a grazing strategy. Important predators include lobsters, crabs, fish and starfish. Populations of urchins were studied in the field by day and night and at different seasons of the year and it was concluded that two forms of predation avoidance were practised: hiding and aggregating. These two behaviours were thought to be mutually exclusive in that hiding is appropriate at low population densities whereas aggregation is only appropriate at high population densities. Further, aggregation only provides effective protection for relatively large urchins. Aquarium experiments showed that small urchins tended to hide whereas large urchins tended to aggregate,

Fig. 5.8. The relation between rate of recession under urchin attack of the edge of a bed of *Laminaria longicruris*, and the biomass of urchins at the edge of the bed. From Breen & Mann (1976).

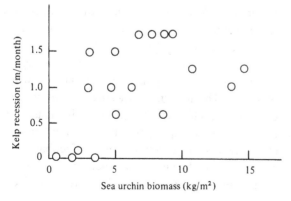

especially in the presence of crab predators. Crabs were found to prefer to attack isolated urchins rather than urchins in aggregations: 'crabs were unable to get their legs and claws around an urchin in an aggregation' (Bernstein *et al.*, 1981). Thus destructive grazing of kelp may be a by-product of the anti-predatory behaviour of urchins at high population densities.

The question is, how have these urchin populations increased to the point at which aggregation, with attendant destructive grazing, becomes the appropriate defensive strategy? Off both California and Nova Scotia the extensive kelp beds of the past suggest relatively recent expansion of the urchin populations. Urchin numbers within normal kelp beds are thought to be controlled by predation: predators shelter under the kelp and urchins hide to avoid being eaten. Breen & Mann (1976) experimentally transferred 400 urchins to a single site within a kelp forest but observed no destructive grazing. The transferred urchins dispersed and, judging from the appearance of empty urchin tests in the area, some of them were eaten. A general reduction in predator numbers, however, should lead to increased numbers of urchins, and this is what is thought to have happened. In 1976 the Nova Scotia lobster populations were estimated to have decreased by 50% over the previous 14 years, the reduction being almost certainly due to fishing. Lobsters are important urchin predators and the reduced predation pressure probably allowed urchin populations to expand. In California it is possible that reduced predation by sea otters may have contributed to the increase in urchins (Estes *et al.*, 1982). Sea otters are voracious urchin predators: a sea otter brings an urchin to the surface and, floating on its back with the urchin resting on its belly, it breaks the urchin open with blows from a stone held between the front paws. Sea otters used to be shot for their fur.

Once urchin populations have increased to the point that aggregation becomes feasible, destructive grazing occurs and reduces the area of the kelp forests. This in turn reduces the habitat for predators and creates barren, urchin-dominated grounds. Thus the switch from low urchin and rich kelp environments to high urchin and low kelp barrens contains an element of positive feedback.

Having destroyed the kelp beds one might expect the urchins to disperse, or starve, and the kelp to regenerate. However, this does not necessarily happen since urchins are generalist feeders and can survive on apparently bare rock by grazing sporelings and detritus (off California detritus from sewage outfalls is thought to be important). They can also absorb dissolved organic compounds through the epidermis, although the importance of this mode of nutrition is not yet clear. Urchins inhabiting barren grounds where kelp used to flourish show decreased growth rates and their gonads are reduced or absent. Populations do, however, persist and off Nova Scotia it appears that as little as 150 g of urchins m^{-2} is sufficient to prevent recolonization by kelp.

Thus the diverse and highly productive kelp forest community and the urchin-dominated barrens appear to be ecologically stable alternative states. The question whether the switch to barrens is irreversible or part of a cyclic change was discussed by Mann (1977) and the problem is how to remove the urchins in the absence of help from predators. In California promising results in regenerating kelp beds were obtained from experiments in which quick lime, a contact urchinicide, was spread on barren grounds. In Nova Scotia disease and mass mortality amongst the urchins has recently led to increased kelp growth (Miller & Colodey, 1983).

In some parts of the world kelp forests do not appear to suffer from urchin attack. In the heavily wave exposed *Macrocystis* beds off Argentina the kelp gribble and wave action were found to be much more important than urchins in causing kelp mortality. Off South Africa patches of *Laminaria pallida* are protected from urchin grazing by wave action which causes the blades of plants at the edges of patches to sweep the surrounding rock. Advancing urchins are deterred or swept away. Further inshore *Ecklonia maxima* forests support populations of urchins which, like *Echinus* around Britain, graze newly settled algae rather than mature kelp.

5.6   The macrofauna

Important macrofaunal elements in kelp forests include large suspension feeders such as mussels, beds of which may carpet rock and holdfasts alike, grazing herbivores such as urchins and abalones, and predators such as lobsters, crabs and starfish. Swimming amongst the fronds are numerous fish (Fig. 5.1).

Apart from the work on urchins, detailed above, little is known of the large invertebrate inhabitants. Lobsters and crabs are important predators of urchins, and lobsters, crabs and starfish all feed on mussels. Abalones may be common in kelp forests and are economically important as food; they eat benthic algae and kelp detritus. Octopus will eat lobsters, crabs and mussels. Most of these species benefit from the shelter of the kelp, but they also occur in rocky environments where kelp is absent. Urchins and starfish may crawl about on the rock surface by both day and night but the other species usually hide under rocks or in crevices during the day and come out to forage at night. They are all, therefore, commoner on bottoms showing high relief – boulders, gullies, ridges, etc. – since these provide more hiding places. The home of an octopus is often advertised by a pile of empty crab and bivalve shells around the entrance. Lobsters and crabs excavate hollows in the loose substrate underneath boulders and pile the soil in heaps outside the entrance, again advertising their presence.

Kelp bed fish have received extensive study in California where 41 species were described as being common in the kelp (Quast, 1971b). Fish populations were sampled by enclosing an area in a ring of netting extending from the surface to the bottom, then spreading poison within the net and collecting the fish by hand (Fig. 5.9). Another method was to lay a line through the kelp and to swim along it recording the fish present within a certain distance either side of the line. Measurements by these two methods led to

Fig. 5.9. Diagram showing a diver applying fish poison within a wall net for quantitative sampling of the fish population. A mixture of poison and sea water is pumped from the dinghy. After Quast (1971b).

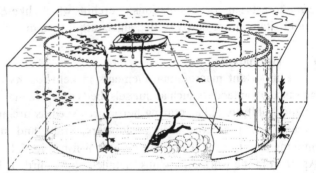

an estimate of fish biomass of 350 kg ha^{-1}, a figure which is similar to the middle of the range of estimates of fish biomass in lakes and coral reefs. The abundance and diversity of fish were positively correlated with bottom relief and depth and were greatest in high relief areas at about 20 m deep. Although fish appeared to be generally more numerous in the kelp than in rocky areas without kelp, all kelp bed species could be found in the latter areas and no species appeared dependent on the kelp. The favourability of the kelp forest as a fish habitat was thought to be related to the way that the kelp, extending up to the sea surface, accentuates the relief of the rocky habitat. This works in two main ways; first the algal surface provides extra living space, and sometimes extra food in the form of epiphytic prey species such as bryozoans and amphipods. Second, the kelp plants provide an extension towards the surface of the normal habitat of the fish.

Quast (1971a) suggested the presence of four zones in the Californian *Macrocystis* beds related to distance from a solid substrate – either sea bed or kelp frond (Fig. 5.10). Zone I, characterized by such fish as blennies, bullheads, eels, gobies and pipe fish, followed the contours of the bottom and extended up kelp columns into the canopy. Most Zone I fish are small and negatively buoyant, lacking

Fig. 5.10. Subdivisions of the giant kelp bed habitat for fish. The columns curving at the top represent the extensions of benthic zones up the kelp fronds. For further explanation see text. From Quast (1971a).

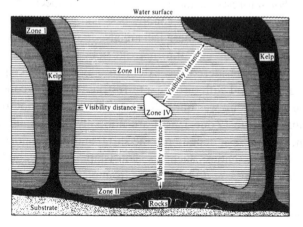

a swim bladder; they rest on the bottom, or amongst algae, or entwine themselves amongst the kelp. Some of them, such as the bullheads and pipe fish, are cryptic in their colour and behaviour. Zone II, a metre or three from the substrate and also extending up the kelp columns; was the province of more conventional looking fish: neutrally buoyant mobile species including various wrasse, perch and bass. These fish foraged actively over the substrate and up the kelp fronds. Zone III was an extension of Zone II, containing similar fish and extending to the limit of underwater visibility away from the bottom and from the columns of kelp. The zone was an open-water habitat, but one from which the substrate (refuge?) was always visible. Quast suggested that the sight of the sea bed, or of kelp columns, constituted a valuable orientational clue and that many species were reluctant to stray, as it were, out of sight of 'land'. It was observed that Zone III was more extensive on days when underwater visibility was good, apparently confirming the importance of sight of the substrate. Zone IV, beyond the limit of underwater visibility, was inhabited by pelagic fish such as barracuda and sprats. These fish occasionally strayed into Zone III.

Californian kelp fish have wide diets with amphipods, crabs, algae, shrimps, polychaete worms, isopods and other fish being amongst the more important food items. Zone I fish took mainly benthic prey whereas Zone III fish often took plankton.

Kelp forest fish have not received detailed study elsewhere in the world. My own observations off the coasts of Britain suggest that in *L. hyperborea* forests the most important fish are species of wrasse and gobies, while above the canopy bass and gadoid fish such as pollack may be seen. All these fish are also found in environments lacking kelp and, as in California, no obligate kelp fish occur. Quast (1971a) considered the extent to which fish in kelp beds could be compared with birds in a terrestrial forest and concluded that the relationship was quite different: most fish do not 'perch', and those that do (Zone I fish) are normally also found on the bottom, and few fish depend on 'the shrub and tree analogues for reproductive activities such as courtship and nestbuilding'. Actually many wrasse do build nests from pieces of algae, but the algae need not be kelp and the nest is usually close to the bottom, often in a rock crevice.

5.7 Energy flow

Energy flow studies have been carried out in Nova Scotia by Mann's group and in South Africa by Field's group. Both show a similar pattern in which most of the primary production is made available to consumers in the form of detritus. The main difference lies in the proportion of live plant eaten by herbivores. In Nova Scotia a substantial amount is directly grazed by sea urchins whereas off South Africa the urchins mainly feed on kelp debris. Fig. 5.11 is an energy flow diagram for St Margaret's Bay, Nova Scotia, and shows the main food chains of seaweed→urchins →lobsters, and detritus and phytoplankton→mussels→starfish. It may also be seen that a substantial proportion of the kelp detritus is exported from the system. Off South Africa the main route appears to be: kelp detritus and phytoplankton→mussels→crawfish; important alternative primary consumers include urchins, holothurians and sponges and top carnivores include cormorants and sea lions.

Fig. 5.11. Energy flow through the bottom community in the kelp forest of St Margaret's Bay, Nova Scotia. Flow is in kcal/m²/y and standing crops in kcal/m². From Miller *et al.* (1971).

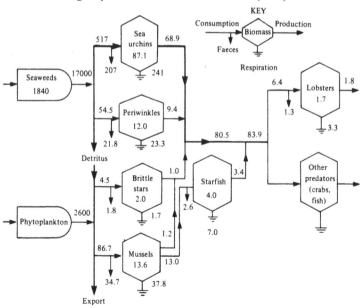

Part III: Coral Reefs

6

Reef structure and environment

6.1   Introduction

Reefs are large solid ridge-like structures rising above the level of the surrounding sea bed and usually reaching close to the sea surface where waves may break on the crest. Coral reefs are those that are built of dead coral skeletons and on which living corals are growing. They are ideal places for diving biological work; the warm clear water adds greatly to the comfort of the diver, making it possible to spend longer underwater and also to dive deeper with greater safety.

Corals are cnidarians; the vast majority belong to the sub-class Hexacorallia, order Scleractinia, but there are some octocorallian and hydrozoan corals. Reef-building or hermatypic corals are the ecological equivalents in the tropics of the temperate kelps since, thanks to the unicellular algae or zooxanthellae which live symbiotically within their cells, they are primary producers (see Chapter 7). A consequence of this symbiosis is that, like kelps, they are limited in their distribution to surface waters where there is adequate light for photosynthesis. Thus, although coral reefs are usually found in clear water, few reef-building corals occur below 40 m and most reef growth takes place in the top 15 m.

In this chapter the aim is to introduce coral reefs as a special type of sublittoral environment. Reef origins are discussed, the physical environment suitable for reef growth is considered, the zonation of a typical reef is described, a personal view of reef growth is presented and a brief account of reef productivity is given. The literature on coral reefs is very large and I have not attempted to cover it here or in the following three chapters which are about special aspects of reef life. For further information the reader should refer to reviews edited by Jones & Endean (1973a, b, 1976, 1977) and to the International Coral Reef Symposia volumes, several papers from which are referred to below.

6.2   Origins

Three types of coral reef are usually recognized: fringing reefs, barrier reefs and atolls. Fringing reefs, so called because they occur as narrow ribbons of reef fringing the coastline, are relatively small structures. The fringing reef crest is normally some tens or few hundreds of metres from the shore and the thickness of the reef from the living surface down to the bedrock or sand on which it rests is usually 5–30 m. These reefs are geologically recent formations: they appear to have grown since the end of the last ice age some 6000 years ago when the melting of the ice raised the sea surface to roughly its present level.

Barrier reefs and the reefs that form atolls are far larger, thicker and older than fringing reefs. Barrier reefs may be several kilometres wide, 100–200 m thick, and may lie many kilometres offshore. Between the barrier reef and the shore the water may be quite deep, up to 100 m. Atolls are ring-shaped reefs, often supporting low-lying ring-shaped islands and always surrounding a more or less circular lagoon. The depth of this central lagoon is related to the size of the atoll: big atolls 30–40 km across have lagoons 60–80 m deep whereas in atolls 5–15 km across the lagoons are 30–40 m deep. Atoll reefs are on the same size scale as barrier

Fig. 6.1. Aerial view of the leeward rim of Eniwetak Atoll. Beyond the short (15–20 m) buttresses the seaward slope drops precipitously into deep water. Behind the buttresses breaking waves indicate a surf zone at the edge of the reef flat (cf. Fig. 6.2). Photograph by J. C. Ogden.

reefs, but some are much thicker: 1000 m or more. Fossil corals retrieved from deep borings to the bases of big atolls show dates going back to the miocene, much earlier than the ice ages, but the corals themselves are typical reef-builders showing that at that time the rock on which they were growing was near the surface. The same applies to barrier reefs which, although not as old as big atolls, are old enough for their origins to be pre-glacial. The explanation, first suggested by Darwin, is that the rocks on which atolls and barrier reefs rest have subsided over the ages (rates of subsidence are in the region of 5–10 cm/1000 years) and upward and seaward reef growth has kept pace with the subsidence to produce the massive reefs observed today. Atolls are thought to have originated as fringing reefs surrounding an emergent volcanic peak which subsequently subsided; barrier reefs probably formed on gradually sinking coastal plateaux. Barrier reefs and atolls are however capped with geologically recent reefs which can be directly compared with fringing reefs (Fig. 6.1).

The importance of a knowledge of the history of reefs lies in the interpretation of structural features of present-day reefs. The recent structures – fringing reefs and their counterparts growing on the tops of barrier reefs and atolls – are the result of reef growth under contemporary conditions: sea level and climate more or less as they are today. Deeper than about 15 m, however, divers encounter structures which are not the result of reef growth but were formed during the ice ages by wind, rain and wave erosion. During the ice ages sea levels dropped to maxima of 100–150 m below the present level, with long stable periods or stands at intermediate sea levels. Large-scale features on the seaward sides of reefs, such as cliffs and terraces, may have resulted from waves undercutting an ice age shoreline. Gullies in the cliffs may mark the courses of ancient streams. These large features – e.g. cliffs on the seaward slope dropping from 15–30 m deep – are not obscured by recent coral growth which forms a relatively thin veneer over the surface.

6.3   Environmental factors

We do not yet know enough about the processes that control reef growth to explain fully why it is that reefs flourish in some places but fail to develop in other, apparently suitable places.

The main environmental factors, however, appear to be light, wave action, turbidity, sedimentation, slope of the substrate, availability of plant nutrients, temperature and salinity. The importance of light has already been mentioned and is explained fully in Chapter 7. Photosynthesis of the algal symbionts promotes the growth rate of coral skeletons and aids coral nutrition, thus adequate light is essential for reef growth.

Wave action is important in washing sea water, with its high oxygen content and stable temperature, over the corals. Planktonic organisms may also be brought in with the waves to aid coral nutrition. The typical distribution of reefs around an island exposed to the trade winds is to find the best developed reefs on the windward side exposed to the greatest wave action. However, this typical distribution is not always found. On the windward sides of Aldabra in the Indian Ocean and of Barbados in the Caribbean the waves wash shorewards over an extensive and shallow sandy platform on which there is little coral growth. It is likely that in these places the sand, caught up by the waves, inhibits coral growth by constant scouring and by showering down as sediment. Further, waves passing over a shallow platform stir fine particles into suspension, making the water turbid. Thus although wave action normally promotes coral growth, it does not always do so.

Turbidity, whether caused by waves, a nearby river or industrial development, is bad for corals since light penetration is reduced and sedimentation promoted. Sediment is an important hazard for corals; apart from the obvious danger of being buried and suffocated by a sudden blanket of sediment (for instance from a river in spate), there is a continuous energy drain to the cleansing mechanisms through the operation of ciliary currents and production of mucus to entangle sediment and remove it from the surface of the colony. Coral diversity and percentage of the reef surface supporting living coral cover are known to be correlated with slope – the steeper the slope, the richer the coral growth. This is thought to be because steep slopes shed sediment more easily than level areas, where sediment may accumulate. River mouths and industrial developments tend also to increase the local levels of plant nutrients, increasing phytoplankton production (so adding to turbidity) and promoting the growth of benthic algae which compete with corals for space on the reef (see Chapter 8). Rivers have the additional disadvantage, from the coral's point of view, of reducing the local salinity. Few cnidarians can tolerate low or variable salinity. Reefs do not usually grow close to river mouths, and those that do are poorly developed and show relatively low coral diversity with populations of just a few hardy species.

The importance of a stable temperature has already been referred to in relation to the favourable effects of exposure to ocean waves. Reef-building corals require stable tropical temperatures for most of the year, but they cannot tolerate very high temperatures such as may occur in lagoons when the trapped water is heated by the sun. Nor, if the reef is exposed at very low tide, can they tolerate for long the heat and drying action of direct sunlight.

6.4   Zonation

The factors that give character to a reef – prevailing wind direction, turbidity, ice age features – are local factors, thus no two reefs are exactly alike: each is unique. The following account of reef zonation is a generalization which should not be expected to apply without modification to all reefs. It mainly relates to windward reefs and is broadly based on my experience in Jamaica and elsewhere, supplemented by various published accounts, especially that of Goreau & Goreau (1973). A plan of the reef zones likely to be found on the north coast of Jamaica is shown in Fig. 6.2. Similar zonation occurs on atolls and barrier reefs except that the details of the seaward slopes may be different and the main lagoons deeper. Atoll islands and barrier reef islands, however, are often separated from the reef flat proper by a shallow lagoon-like channel corresponding to the lagoon behind a fringing reef.

6.4.1.   Lagoons

Lagoons may be wide, narrow, shallow or deep: rough limits might be 5 m to several kilometres wide and 0.5 to 50 m deep. They are sheltered from wave action by the reef and are therefore calm. The bottom is sandy or silty and may be colonized by algae and sea grass. A number of factors militate against coral growth in lagoons: these include sedimentation, variable temperature, variable oxygen tension and little planktonic food. All these

adverse factors result from the sheltered, enclosed and shallow
nature of lagoons: when a break occurs in the reef, allowing fresh
sea water through, coral growth within the lagoon is stimulated.
Despite the adverse conditions, however, corals do grow in
lagoons, usually in isolated hillocks with low species diversity.

Fig. 6.2. Diagrammatic plan of reef zones based on the
characteristic features of the fringing reef on the north coast of
Jamaica. The figures on the right indicate the approximate
depths (m) of the zones. For further details see text.

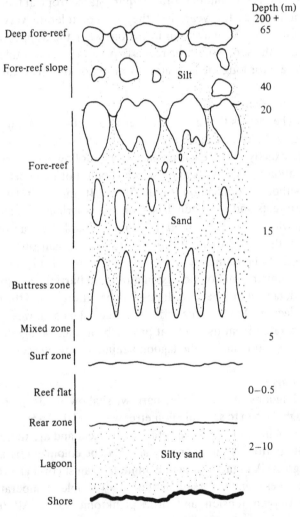

Depth (m)

6.4.2 The reef crest

At its seaward edge, the lagoon floor rises to meet the reef flat. This is a shallow (0.5–1 m), often extensive area, sometimes exposed at low tide, and characterized by water movement which increases in turbulence as one proceeds seawards. Islands of broken coral are sometimes thrown up on the reef flat by storms. The reef flat is bordered at its inner edge by a rear zone, the seaward margin of the lagoon; some coral growth occurs here stimulated by water driven over the reef by waves. The true reef flat is composed of reef framework and rubble cemented by encrusting coralline algae. Rather little coral growth occurs here but colonial anemones (zoanthids) and algal turfs may cover large areas. (One reef flat in Jamaica earned the name of the dunlopillow zone because of the thick carpet of *Zoanthus* which occurred there and was so comfortable to walk on.) Seaward from the reef flat is a surf or breaker zone which, as its name implies, is the most turbulent zone on the reef. Corals of the genus *Acropora* are characteristic here and the hydrozoan stinging coral *Millepora* may be important. In the surf zone on very exposed reefs, especially in the Pacific, a calcareous algal ridge sometimes develops. These ridges, built of the skeletons of encrusting coralline algae, are dissected by surge channels through which the water seethes as the waves break and retreat. In the Caribbean the large tree-like elk-horn coral *Acropora palmata* (Fig. 6.3) is characteristic of the surf zone and sometimes occurs in extensive monospecific stands. Seaward of the surf zone there may be a narrow moat containing dead coral, broken from the surf zone by storms. Coral diversity on the reef crest is usually low, probably because of the severe physical stresses associated with life on this part of the reef.

6.4.3 The buttress zone

Seaward of the surf zone and decending from 2 or 3 m to 5 or 6 m is a zone of active coral growth and fairly high diversity known as the mixed zone. This merges into the most diverse zone of all and the zone of most active coral growth, the buttress zone. Buttresses are massive finger-like spurs jutting out to sea and clothed in living coral. They may be 30 m or more long and 3–12 m high and are most spectacular structures (Fig. 6.4). Between the

Fig. 6.3. A stand of elk-horn coral *Acropora palmata* in an area
exposed to wave action from left to right of the photograph.
Branches are about 1 m long. 3 m, Tobago, Caribbean.

Fig. 6.4. Scenic view of the buttress zone on a Jamaican north
coast reef at about 10 m deep. Based on a sketch by P. D.
Goreau, after Goreau & Goreau (1973).

buttresses are steep-sided canyons floored with rubble and sand. Buttresses and canyons (or spurs and grooves) do not appear to be erosional phenomena since erosion does not seem to occur in the canyons, and the buttresses, on the contrary, appear to be growing both upwards and seawards. Goreau & Goreau's (1973) interpretation was that the canyons are drains for sediment and rubble originating higher up on the reef. Sediments take the form of calcareous algal sands, coral sediments (produced by coral grazers and boring organisms), calcareous debris originating as the skeletons of barnacles, molluscs, echinoderms, etc. and finely divided organic matter such as coral mucus, faecal pellets and plant detritus. This material moves down the canyons under the influences of gravity and wave-induced seaward flowing currents and inhibits coral growth. On the buttresses, however, growth is promoted since the steep slopes exposed to turbulent water easily shed sediment. Thus as soon as a sedimentary chute develops – and they must develop since reefs produce vast quantities of sediment (see below, §6.5) – the subsequent growth of the corals ensures that it persists and perhaps develops into a canyon. Unlike river valleys, therefore, canyons develop by the upward growth of their surroundings. A fate that may overtake canyons is to become roofed over by vigorous lateral growth of corals from the tops of the buttresses. The tunnels delving into the reef which result from this process are great fun to swim through. They open from time to time through chimneys onto the surface of the reef and sometimes extend right back to open on the reef flat providing good evidence for the seaward growth of the whole reef.

The framework of the buttresses is built from large corals, the skeletons of which are buried *in situ* by further coral growth and addition of calcareous sediments. Goreau & Goreau (1973) called these framework builders 'primary hermatypes' and they include on Caribbean reefs the large branched *Acropora palmata,* the brain coral *Diploria* and, perhaps most important on the buttresses, the massive coral *Montastrea annularis.* The framework is filled in by secondary hermatypes: smaller corals and calcareous algae which are not usually buried *in situ* but which, on many reefs, actually contribute more calcareous material to the reef than do the primary hermatypes. Much of this material is cemented by encrustations of coralline algae, foraminiferans and the stinging coral *Millepora*, but the reef remains a fairly loose and porous structure. The buttress zone ends at about 15 m or shallower with the buttress tips often resting on sand.

6.4.4   The seaward slope

Beyond the buttresses lies the seaward slope, the extent of which depends on the slope of the underlying rock. On the north coast of Jamaica the seaward slope, from about 15 m to about 70 m is often less than 100 m wide and parts of it are precipitous. Two 'drop-offs' usually occur: the first, at 15–20 m, is the rim of an ice age cliff which drops to about 40 m (Fig. 6.5). Below this is a steep muddy slope with coral outcrops which descends to the second drop-off at 55–70 m (Fig. 6.6). This is the edge of a vertical cliff that plunges into the depths. Three zones can be distinguished on the seaward slopes of Jamaican reefs: the fore-reef, the fore-reef slope and the deep fore-reef. The fore-reef is a fairly rich zone of coral growth with fairly high coral diversity which extends in patches and hillocks (and often wide areas of sand and sea grass) from the tips of the buttresses to the edge of the first drop-off. Here massive clumps of coral have accumulated to form battlements on the edge of the cliff. These clumps look not unlike truncated buttresses and there are sediment chutes between them down which fine material drains (Fig. 6.5). The coral growth on the drop-off may extend some way down the cliff in overhanging masses which can break off and slump onto the muddy slope below to form coral pinnacles (Fig. 6.7). At the base of the cliff the muddy fore-reef slope extends down to the second drop-off. Evidence of sediment transport down the slope is common and in places tiny local avalanches can easily be started. Pinnacles and smaller blocks of coral from above, as well as rocky outcrops, project from the mud and support a fairly rich coral community but with a reduced species diversity; the genus *Agaricia* is characteristic of this zone on Caribbean reefs. Also present are algae, gorgonians, black corals and sponges, the last two becoming increasingly common with depth. At the second drop-off another rocky sill with sediment chutes is usually present and is tenanted by massive sponges. Over the drop-off lies the deep fore-reef which is not a reef at all but a near-vertical escarpment in which the importance of reef building corals is greatly diminished (Fig. 6.8). Investigation of this zone by divers is barely practicable because of the short time available at the site, the long decompression necessary and the dangers of nitrogen narcosis.

Fig. 6.5. Scenic view of the fore-reef and fore-reef slope on a Jamaican north coast reef. Buttress tips are seen at the top left. The drop-off is from about 20 m to 40 m and sediment drains down gullies on the face of the escarpment. Based on a sketch by P. D. Goreau, after Goreau & Goreau (1973).

6.4.5 *Leeward reefs*

So far only windward reefs have been considered. Zonation on leeward reefs is much simpler since the lack of continuous wave action eliminates many of the windward zones. The lagoon is usually narrower, the reef flat, which probably results to a large extent from storm damage, is less well developed, the surf zone is absent and the buttresses are truncated or missing. A mixed zone, therefore, leads directly from a poorly developed reef flat

Fig. 6.6. Scenic view of the fore-reef slope and deep fore-reef. The fore-reef drop-off to about 40 m is at the top left. The diver at the second drop-off is at about 70 m. Based on a sketch by T. F. Goreau, after Goreau & Goreau (1973).

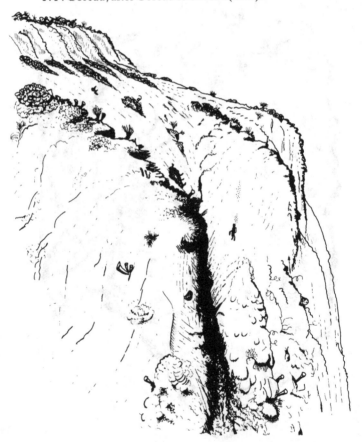

down to the fore-reef and fore-reef slope. In addition to the lack of wave action, leeward reefs may be affected by turbidity and sedimentation since detritus and silt, washed from windward areas, may accumulate to leeward.

The transition from exposed windward reefs to sheltered leeward reefs is also accompanied by a change in the species composition and distribution of the coral communities. Corals which, on windward reefs, are characteristic of the most turbulent zones, may be absent on leeward reefs and their places taken by species which, on windward reefs, are found in deeper water or in more sheltered back reef positions. This transition through several levels of exposure is illustrated in Fig. 6.9 and shows the great importance of wave action as an environmental factor.

6.5   Reef growth

Implicit in the above account is the assumption that reefs grow towards the surface and out to sea by the accumulation of calcium carbonate skeletons of organisms. An average growth rate

Fig. 6.7. Possible stages in the formation of coral pinnacles on the fore-reef slope. (*a*) the drop-off from 20–40 m (probably an ice-age cliff, the shape of which is inferred) with a thin covering of coral, and some rubble on the slope below; (*b*) overhanging coral growth with blocks on the slope below; (*c*) pinnacles formed by slump of overhanging coral. After Goreau & Goreau (1973).

     (*a*)   (*b*)   (*c*)

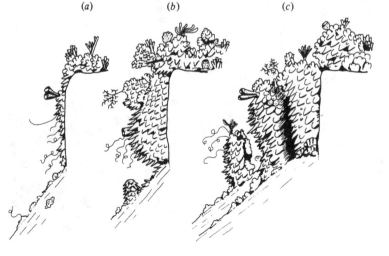

Fig. 6.8. The deep fore-reef on the north coast of Jamaica. The wire-like growths are black corals (Antipatharia), also present are gorgonians, large sponges and sediment. Photograph by T. F. Goreau. From Goreau & Goreau (1973).

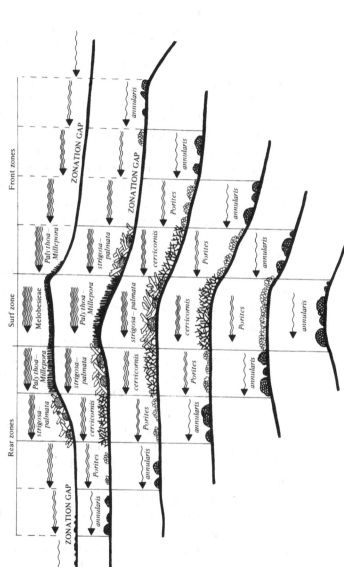

Fig. 6.9. Diagram illustrating the effect of degree of exposure to wave action on the species composition and distribution of characteristic dominant organisms in Caribbean shallow reef communities. Relative degrees of wave action are indicated by the number of tails on the arrows. The most exposed reefs are characterized by encrusting coralline algae (Melobesieae) in the surf zone, and with decreasing exposure to wave action the characteristic surf zone species are as follows: the colonial zoanthid *Palythoa* and the stinging coral *Millepora*, the brain coral *Diploria strigosa* and the elk horn coral *Acropora palmata*, the stag horn coral *Acropora cervicornis*, the finger coral *Porites furcata*, and on the most sheltered reefs the massive coral *Montastrea annularis*. 'Zonation gaps' on the exposed reefs probably result from scouring of these regions during storms. After Geister (1977).

of 1 mm year^{-1} is often cited. No reef growth can take place above sea level so islands made of reef material either result from upward movement of the Earth's crust, or a drop in sea level, or from storms. Upward growth of reefs stops at about the low tide mark, but if the land is sinking or the sea level is rising, reef growth can keep pace and the resulting accumulation, as in atolls and barrier reefs, may become impressively massive. Coral reefs are composed of big coral skeletons (framework) cemented together by encrustations and with the spaces filled by smaller pieces of skeleton and calcareous sediments. This loose structure is further consolidated to form reef rock by the growth in the remaining interstices of tiny calcareous crystals of micrite cement which fill and bind the porous skeletons and sediment. Active coral growth producing reef framework mainly occurs at the seaward edges of reefs: in the surf, mixed and buttress zones. Here one can imagine that the mixed zone gradually grows towards the surface to become a surf zone while the original surf zone becomes incorporated in the reef flat. On the buttresses upward coral growth also occurs but seaward growth can only take place when a hard substrate is present at the outer ends of buttresses. One can imagine that hard substrates are provided here by the seaward collapse of overhanging coral colonies growing at the ends of the buttresses. The development of coral patches on the fore-reef may also provide hard substrates for the continued seaward growth of buttresses. On the fore-reef slope coral growth is unlikely to contribute to the growth of the main reef since transport of skeletons is mostly downwards. However, reef features such as pinnacles may be formed (Fig. 6.7).

The above ideas leave out an important environmental factor: the occurrence of violent cyclonic storms, typhoons and hurricanes. Although in any one place these storms are infrequent, coral growth is so slow (see Chapter 7) that the changes produced by a single storm may be equal to the changes produced by many years of coral growth. The well known reef at Discovery Bay on the north coast of Jamaica received a pounding from heavy seas as hurricane Allen passed by 50 km to the north during August 1980. Divers reported massive destruction of coral, especially in the surf zone but extending as deep as 50 m (Fig. 6.10). Large framework builders such as *Acropora palmata* had been overturned, broken and reduced to rubble, some of which had been thrown up onto the reef flat. It is likely that the reef flat extended slightly seawards as a result of the storm, and on the buttresses the overturning of large colonies could have provided further hard substrate for extra seaward growth. Thus although storms are destructive to corals, they may be constructive agents in reef growth. They may even be constructive in terms of coral reproduction since not all fragments of broken colonies are killed; some regenerate in their new locations and form new colonies. Recovery from storm damage depends on the severity of destruction and may take anything from 5–50 years (Pearson, 1981).

No account of reef growth would be complete without reference to bioerosion. Boring organisms – algae, sponges, polychaete worms, bivalve molluscs, etc. – are constantly eroding the corals and thus the entire reef. Many fish and sea urchins graze the reef surfaces, converting coral skeleton and reef rock to sand. These interactions are discussed in Chapter 8 but it is worth looking here at the quantities involved since, if the sediment produced by

Fig. 6.10. Storm damage to an elk-horn coral *Acropora palmata* community. An overturned stand is distinguishable at centre-right; compare with Fig. 6.3. Some regeneration is visible at centre-left. The fish in the foreground is about 10 cm long. 3 m, Tobago, Caribbean.

bioerosion is removed from the reef, then in order for the reef to grow, calcification must exceed bioerosion. Measurements of reef calcification in shallow water (0–10 m) range from about 4–15 kg $CaCO_3 m^{-2} yr^{-1}$. This is almost entirely due to coral growth; coral-line algae *in situ* add relatively little. The amount of bioerosion by sessile boring organisms, mainly sponges, depends on the depth and the percentage of the reef rock surface covered by living coral. Bak (1976) estimated that corals growing in shallow water off Curaçao lost a mere 1–2% of their annual weight increment by boring. However, corals living deeper on the reef often grow more slowly and colonies have been found in which 25–30% of the skeleton has been removed by boring. Borers usually only attack exposed reef rock and the exposed skeleton at the bases of coral colonies, thus the higher the cover of living coral and of other encrusting organisms, the less room for borers. The relative pro-portions of coral cover and exposed reef rock vary greatly from reef to reef: coral cover is often between 40 and 80%. On a shallow leeward reef in Barbados Stearn *et al.* (1977) found about 40% coral cover and 20% bare reef rock. They calculated that boring bioerosion removed only 0.14 kg $CaCO_3 m^{-2} yr^{-1}$ from the corals but 2.2 kg $CaCO_3 m^{-2} yr^{-1}$ from the reef rock (Stearn & Scoffin, 1977). Even higher are some estimates of sponge bioerosion in Jamaica: up to 7 kg $CaCO_3 m^{-2} yr^{-1}$. Grazing bioerosion by fish and urchins may be even more serious. Estimates vary from 0.1 kg $CaCO_3 m^{-2} yr^{-1}$ for fish (Lewis, 1977) to 15 kg $CaCO_3 m^{-2} yr^{-1}$ for urchins on the Barbadian reef mentioned above (Stearn & Scoffin, 1977). Calcification on this reef is also 15 kg $CaCO_3 m^{-2} yr^{-1}$ so, if the activities of borers and fish are included, it appears that without sedimentary deposition and consolidation this reef would be getting smaller rather than growing.

More measurements of grazing and boring rates are required for the importance of bioerosion to be fully understood, but if bio-erosion is normally about the same as biological calcification, net reef growth would seem to depend more upon the processes of sedimen-tation and cementation than on coral growth *per se*. Corals may merely provide a favourable environment for sediment accumu-lation. This possibility was raised by Davies (1977) with particular reference to the Great Barrier Reef province. Divers here have found that carbonate accumulation is occurring more rapidly to leeward than to windward; indeed it appears that windward carbonate production is mostly transported to leeward in the form of sediments and rubble. Here secondary coral growth and patch reef development are taking place and lagoons are gradually being filled in. Davies concluded that these reefs were growing backwards and suggested that forward growth at the buttresses may be the exception rather than the rule.

6.6   Reef productivity

It seems appropriate to end this chapter with a brief account of the productivity of reefs. Coral reefs teem with life – they are highly productive – but this has always been a bit of a puzzle because many of the richest reefs are in geographical areas noted for their low productivity. Tropical oceans tend to have low concentrations of plant nutrients in their surface waters and thus support relatively low levels of primary production. In these watery deserts coral reefs are surprising oases of flourishing life.

Reef productivity has usually been estimated as a whole by measuring the increase in oxygen concentration in the water flowing across a reef (Lewis, 1977). It has been found that oxygen concentration increases between upstream and downstream sampling points during the day, and decreases at night (Fig. 6.11). The increase by day is attributed to net community primary production and the night decrease to community respiration. Community respiration during the day can be measured by darkening the flow path (e.g. by building a black plastic tunnel) or by estimating from the night measurements. Gross primary production can then be obtained by adding the daytime net primary production to the daytime respiration. Over the dozen or so reefs where such measurements have been taken the gross primary production has been found to range between 0.3 and 5 kg C m^{-2} yr^{-1}, and the ratio of gross primary production to total community respiration (P/R) to range from 0.6 to 2.6 with most estimates lying between 1 and 1.5 (Lewis, 1977). Gross primary production on reefs is usually at least an order of magnitude greater than in the surrounding ocean water.

These measurements of reef productivity can be criticized from a number of angles, perhaps the most important of which is that most measurements were taken, for the sake of convenience, in shallow water on the reef flat. Exchange of oxygen between water and air during passage of the water from upstream to downstream sampling points is the most obvious hazard in shallow water and could lead to underestimates of both production and respiration. Another hazard lies in the use of measurements of reef flat production to estimate production in other parts of the reef. Reef flats are usually heavily colonized by algae and fewer corals occur here than deeper on the reef; also light intensity is much greater in shallow water. Thus the reef flat is probably the most productive part of the reef and to extend these measurements to describe the whole reef is likely to produce a considerable overestimate. In a study in deeper water (4 m) on a Puerto Rican reef, using divers to set up the flow path (Fig. 6.12) and take the water samples, Rogers (1979) found gross primary production to be $5.4 \, g \, O_2 \, m^{-2} \, day^{-1}$ (roughly equivalent to $1.4 \, kg \, C \, m^{-2} \, yr^{-1}$) but found a 24 h P/R of only 0.7 indicating net consumption at this depth.

Fig. 6.11. Diurnal curves of oxygen concentration in water flowing over a coral reef. Values represented by the dotted line were recorded at the outer edge of the reef flat while the solid line describes oxygen levels in the rear zone (see Fig. 6.2.). The gap between the lines thus represents the oxygen produced or consumed during passage of water across the reef flat. From Lewis (1977).

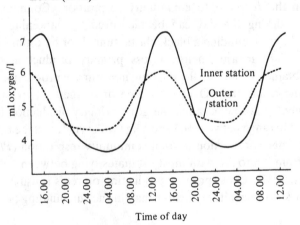

Thus it seems likely that net production occurs in shallow water where there is abundant algal growth, and net consumption in the deeper parts of the reef. Algae rather than corals are probably the chief primary producers. These algae produce detritus through mechanical damage and are grazed by fish, urchins and small crustaceans (see chapters 8 & 9) which generate detritus in their faeces. At Eniwetak in the Marshall Islands it was estimated that fish grazed 60% of algal production but assimilated only about 10% of what they ate; the rest was defaecated as detritus. Corals also produce detritus, mainly in the form of mucus. Detritus, together with attendant bacteria, is consumed by suspension and deposit feeders and this pathway of energy flow is probably just as important on reefs as it is elsewhere in the sublittoral. P/R estimates close to unity, and below unity in deeper water, suggest that reefs consume most of what they produce and thus may not export much energy to the surrounding ocean.

Finally we return to the inorganic nutrients necessary to keep the high production going. The main nutrients, phosphate and nitrate, are almost non-existent as free ions in the ocean waters but are available in small amounts in the plankton. As water crosses a reef zooplankton density decreases by 20–60% presumably as a result of predation by reef organisms, particularly corals. This represents a small input of nutrients which must be of particular benefit to the zooxanthellae within the corals. A further source of nitrate on the

Fig. 6.12. Diagrammatic representation of an underwater metabolism channel 10 m long by 2 m wide by 1 m high, set up parallel to a long-shore current. Water is sampled at upstream and downstream ends to monitor oxygen production or consumption. After Rogers (1979).

reef is the activity of nitrogen-fixing organisms, mainly blue-green algae, which increase the nitrate content of the water as it flows across the reef. No extra source of phosphate is known, however, and the only possible explanation is to suggest that the phosphate within the system is conserved on the reef by tight and rapid recycling (Lewis, 1977).

7

Nutrition and growth of reef corals

7.1   Introduction
A major feature of cnidarians is the presence of nematocysts on their tentacles. This suggests that the group as a whole is
carnivorous. However, reef-building corals, and some representa-
tives of other cnidarian groups on coral reefs such as anemones,
zoanthids and gorgonians, contain single-celled plants or zooxanthellae in their tissues, as well as nematocysts on their tentacles. The
zooxanthellae carry out photosynthesis, which presumably is of
some benefit to their hosts. If the benefit is nutritional, why do the
corals need nematocysts? Many workers have demonstrated carnivorous behaviour in corals, and more recently others have shown
that products of photosynthesis by the zooxanthellae find their way
into the surrounding coral tissues. Thus carnivorous and photosynthetic nutrition complement each other. Different corals,
however, set different balances between the two modes of nutrition
and this appears to correlate with factors, such as distribution on
the reef and shape of colony, which are visible to divers. Experiments and observations on coral nutrition have been carried out
both in the laboratory and on the sea bed, the latter being the preferred environment since there is minimal disturbance to the coral
and conditions are as natural as possible.

A second benefit which zooxanthellae confer on their hosts is to
speed up the rate of calcification, and so of skeleton growth. Corals
grow more quickly in the light than they do in the dark. This effect
is thought to occur partly by photosynthesis removing the bypro-
ducts of the calcification reaction:

$$Ca(HCO_3)_2 \rightleftharpoons CaCO_3 + H_2O + CO_2$$

Accumulation of water and carbon dioxide would slow down calcification and their removal should promote it. The effect of light

upon the growth of corals is another aspect of reef ecology which is visible to divers.

This chapter on nutrition and growth is therefore in large part an account of the importance of light in the lives of reef corals.

7.2   Nutrition

Starting with primary production, several workers have demonstrated that reef-building corals in well-illuminated surface waters produce more oxygen by photosynthesis than they consume in respiration. *P/R* is almost always positive, is often about 2, and is sometimes as high as 4–5. However, not all corals live close to the surface and it has also been found that photosynthetic capability varies according to the depth at which the corals are living. It is of course expected that photosynthesis decreases with increasing depth because of the diminishing availability of light, and this has been demonstrated. The tendency is partly offset, however, by an adaptation of the photosynthetic system to lower light regimes in which the light intensity at which the system becomes saturated decreases with increasing depth – in other words photosynthesis becomes more efficient. (Similar adaptations have been observed in understorey plants in terrestrial forests.) A further supplement to net photosynthesis at depth results from a reduced respiration rate in deeper colonies. Table 7.1 shows the respiratory and photosynthetic performances of colonies of the Caribbean coral

Table 7.1 *Respiration and photosynthesis in colonies of* Montastrea annularis *from various depths*

Depth (m)	Respiration rate ($\mu l\, O_2\, cm^{-2}\, h^{-1}$)	Photo-synthesis saturation level (1x)	Duration of saturation (h day^{-1})	Carbon fixed (g m^{-2} day^{-1})	% of 24 h require-ment
2	18.05 ± 7.2	6996 ± 538	10.5	2.94	108
10	12.25 ± 2.5	4779 ± 161	9.5	1.41	90
20	14.13 ± 3.2	4198 ± 254	6	1.29	73
40	10.34 ± 1.4	3724 ± 344	0	0.59	45

After Davies (1977)

Montastrea annularis. This coral grows in the form of massive rounded 'heads' (Fig. 7.3); it is common in the buttress zone where it is an important primary hermatype but it also commonly occurs in deeper water on the fore-reef slope. The adaptation to depth of both the photosynthetic saturation level and the respiratory rate may be seen in the table; in addition the reduction in light intensity with depth is evident in the figures for the number of hours per day at which saturating intensities are experienced at the various depths. Finally Table 7.1 shows that if photosynthesis and respiration are added up over a 24 hour cycle, then only in shallow water is it possible for this species to support itself by photosynthesis. (An assumption here is that all net primary production of the zooxanthellae is transferred in a usable form to the coral tissues.) In deeper water, despite its adaptations, it needs an additional means of nutrition to survive.

The data in Table 7.1 were the results of relatively short-term laboratory experiments coupled with careful collection of colonies and *in situ* light measurements. An alternative approach is to provide light as the only source of nutrition and to look at survival and growth in the long-term. Fig. 7.1 shows the results of an experiment of this type in which the possibility of nutrition by particle capture was removed by filtering the water in which the corals were maintained. The most likely interpretation of these results is that photosynthesis of the zooxanthellae supported the growth of the corals. An alternative possibility is that growth was supported by absorption of dissolved organic matter. A field demonstration that zooxanthellae can support coral growth in shallow water was provided by an observation made on a dive by Porter (1974) who found a colony of *Porites furcata* growing in the centre of the osculum of a 1.5 m diameter barrel sponge, bathed in biologically filtered water.

The principal alternatives to photosynthesis as modes of nutrition in corals are the capture by tentacles and nematocysts of zooplankton and the capture by entanglement in mucus of smaller, less active particles such as detritus and bacteria. These modes of feeding have been observed both in the laboratory and in the field. Starting with zooplankton capture, it is well known that most corals only expand their polyps and extend their tentacles at night, when reef plankton is most abundant. One of the attractions of night-diving on coral reefs is the sight of the expanded polyps; however, if one stops to admire closely, the extended tentacles are likely to be obscured by swarms of plankton attracted to the diving lights. *Montastrea cavernosa* (Fig. 7.2) is a Caribbean coral with particularly large polyps, so large that Porter (1974) was able to poke the tip of a hypodermic syringe into the mouths of polyps and extract the gut contents. Porter used this technique in the field to sample a large number of polyps two hours after dusk on a reef off Panama.

Fig. 7.1. Growth of corals in the laboratory in filtered and unfiltered sea water. Points are the means, with ranges, of the number of specimens in brackets. From Johannes (1974).

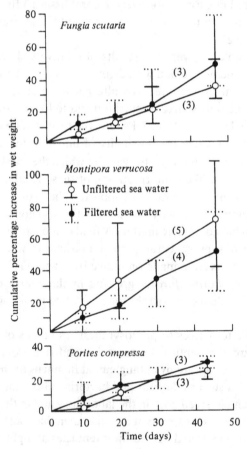

Fig. 7.2. Close-up photographs of the massive Caribbean coral *Montastrea cavernosa*. (*a*), during the day with polyps closed; (*b*), at night with polyps open

Analysing the gut contents, he found that usually one-third to one-half of the sampled polyps had captured one or more zooplankters; a wide variety of prey items was found, the commonest being copepods and crustacean larvae. The assemblage was unlike that captured over the reef in a plankton net and was thought to consist largely of bottom living forms which rise from the reef at night. Porter estimated the calorific value per polyp of the captured prey and calculated that the two hour feeding period from dusk to sampling time yielded 0.2–11.0% of the daily nutritional requirements of the coral. Thus even if the observed rate of feeding was maintained throughout the night (which is unlikely since the zooplankton probably rises to the surface out of reach of the tentacles and does not return to the reef until dawn) the maximum contribution to nutrition is only about 60% of daily requirements.

Another field study on carnivorous behaviour was that of Johannes & Tepley (1974) on *Porites lobata* in Hawaii. This species has rather small polyps (about 1.3 mm in diameter when expanded) which, unlike most corals but like other members of the genus *Porites*, are expanded by day as well as by night. Time-lapse underwater photography was used – every 6–7 seconds for up to 24 hours – to assess the number of feeding polyps in a colony over the 24 hour cycle. Feeding polyps were recognized as those which showed a characteristic contraction, this criterion being established by prior laboratory work. Mean feeding rate was found to be about 3% of polyps feeding at any one time, and was slightly more frequent (4%) at dawn and dusk than at other times. Using the most optimistic assumptions it was calculated that this feeding rate could not support more than 20% of the daily requirements of the animal tissue of the coral, and thus was not an important source of energy for the species.

Feeding on small particles using mucous nets and strings has been studied both in the field and in the laboratory by Lewis & Price (1975). They found that the oral discs of the polyps could secrete mucus which was carried away from the mouth in sheets or filaments by ciliary currents. Small particles became entangled in the mucus which accumulated in strings between the polyps. These particle-laden strings can be ingested since they are attached to the mouth where powerful inwardly-directed, ciliary currents can reel them in. Water movements close to the polyps increased the efficiency of particle capture by lifting the mucous net from the oral disc and wafting it about. This mucous feeding took place by both day and night with either open or closed polyps. Lewis (1976) investigated the rate of mucous feeding in a number of coral species in the laboratory and found a range of 16–145 ml water cleared of particles per cm^2 of living coral surface per hour. The commonest clearance rate in a standard experiment with a magnetic stirrer producing a current of about 1.4 cm s^{-1} was about 100 ml cm^{-2} h^{-1}. Clearance rate was greatly increased by increasing current speed. There is sufficient particulate organic matter in sea water for mucous feeding to make a substantial contribution to coral nutrition, assuming that the corals can digest the detritus that makes up the bulk of the particles.

Lewis & Price (1975), on the basis of their observations, divided corals up into three feeding groups: those mostly using tentacular capture, those that mostly use mucous strings, and those using a combination of both methods. The last group contained the majority of the investigated species. However, Lewis & Price did not consider the input from the zooxanthellae which, as detailed above, may often be the most important source of nutrition. Thus, although they cited several species of *Porites* as feeding mainly by tentacular capture, species of this genus are more likely to be primarily photosynthetic in their nutrition: it was a colony of *P. furcata* which Porter (1974) found growing in a barrel sponge and it was *P. lobata* in which Johannes & Tepley (1974) found such a modest rate of tentacular feeding. Indeed, Johannes & Tepley (1974) observed that *P. lobata* was very inefficient at capturing zooplankton – even tiny copepods 0.3 mm long were observed to escape after bumping into the tentacles. Their description of tentacle feeding in this species contrasts strongly with Porter's (1974) description of feeding in *Montastrea cavernosa*, the powerful tentacles of which can capture and subdue quite large prey such as errant polychaetes 7 mm long.

Porter (1976) suggested that the size of the polyps and the surface area to volume ratio of the colony provided a guide to the relative contributions of photosynthesis and tentacle capture in the nutrition of the species. Small polyps, it was suggested, are inefficient zooplankton traps, but a high surface area is necessary for harvesting light. Porter incorporated these ideas into a model of resource partitioning amongst the coral community on a reef by suggesting the presence of a branched, small polyped, light-harvesting canopy composed of such corals as *Acropora* spp. and *Porites furcata*, overtopping an understorey of more massive, large polyped, carnivorous corals such as *Montastrea* and *Diploria*. This is an attractive model, but requires more work to develop it. Indeed the nutrition of corals is clearly not merely a species-specific matter with each species having a characteristic balance between the various sources of food: it is also a colony and time-specific matter since the availability of all nutritional sources will vary independently according to such factors as location on the reef, time of year, and weather. The general nutritional strategy of corals is therefore of necessity somewhat opportunistic with three different but complementary methods of nutrition and a diversity of environments in time and space in which to deploy them.

7.3   Growth rates

Biologists have been measuring coral growth for at least 100 years (Buddemeier & Kinzie, 1976), interest being initially stimulated partly by the biological fascination of a plant-like animal growing into a stone, and partly by the idea that coral growth might be proportional to reef growth. We now know that growth rates *per se* are ample for reef growth, but that, as explained above (§6.5), the multiplicity of other influences invalidate a simple extension of coral growth to reef growth. The biological fascination of the problem has remained, however, and has been supplemented in recent years by an interest in the mechanisms of calcification. The advent of diving permitted the extension of study into deeper water and has also stimulated an ecological view of coral growth, in particular the relationship between growth forms and environment.

The methods of measuring coral growth are as various as the reasons for study. The simplest, direct and most widely used method is the increase in linear dimensions with time. Study by this method had shown a huge diversity of growth rates both within and between species (Buddemeier & Kinzie, 1976) but the broad generalization that branching corals grow faster than massive corals is usually justified. Representative rates in shallow water for branching corals of the genus *Acropora* are 100–200 mm yr^{-1}, contrasting with typical shallow water rates for the massive coral *Montastrea* of 6–12 mm yr^{-1}. These measures are of increase in individual branch length in *Acropora* and increase in radial dimension (usually upwards) in *Montastrea*. The genus *Acropora* contains some of the fastest growing corals, *Montastrea* is more typical; with the exception of *Acropora*, shallow water growth in the region of 2–20 mm yr^{-1} covers most species.

A limitation of measuring growth by linear increments is that corals have such a diversity of shapes (Fig. 7.3). Massive corals have a low surface to volume ratio compared with branching and foliaceous forms. The surface is the living animal, the volume is

Fig. 7.3. Various coral growth forms: (*a*) open branching – staghorn coral *Acropora cervicornis*, photograph by A. R. Ainslie; (*b*), (*c*), (*d*) & (*e*) overleaf.

Fig. 7.3. Growth forms: (*b*) densely branching – *Madracis mira-bilis*; (*c*) massive heads – *Montastrea annularis*;

(*d*) massive hemisphere – brain coral *Diploria*; (*e*) plate-like –
Agaricia.

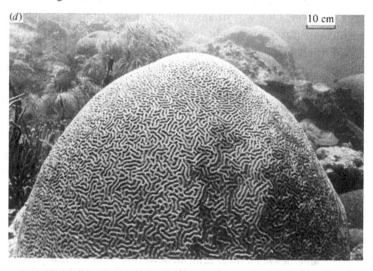

dead skeleton, thus branching corals, because of their greater surface, are 'bigger' than massive corals and so should be expected to grow faster. One way of answering this objection is to measure calcification rates by weight increments. This was done by Bak (1976) who weighed colonies of several species of coral at monthly intervals for up to two years. Weighing was carried out underwater, the standard error of the method being ± 0.1 g. His findings support the contention that growth rate is proportional to surface area. Fig. 7.4 shows the results for three contrasting species. *Acropora palmata*, a species with thick, flattened branches, was found to grow most rapidly, with the more delicately branched *Madracis mirabilis* and the massive *Montastrea annularis* coming roughly equal second. It may be seen that in all three species the

Fig. 7.4. Growth of colonies of three species of coral over a two year period at various depths: *Acropora palmata* at 3 m, *Madracis mirabilis* at 12 m, *Montastrea annularis* at 15 m. (*a*) the cumulative increase beyond the initial weight, in g dry weight; (*b*) the growth rate in g/30 days. Initial weights of colonies were about 100 g (*Acropora*), 150 g (*Madracis*) and 1330 g (*Montastrea*). Curves were calculated from monthly measurements on about 12 colonies in each species. Modified from Bak (1976).

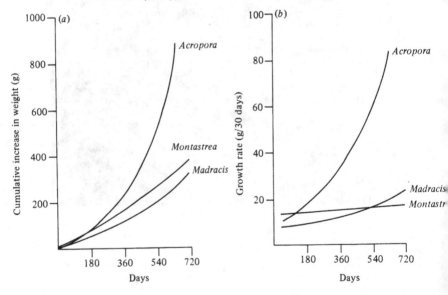

growth increment increases with increasing size, but that this increase is much more marked in the branched species, indicating the importance of surface area. In fact at the end of the experimental period *Madracis* was growing faster than *Montastrea*. However, the initial weights of the colonies are relevant here: the *Acropora* specimens started at about 100 g, *Madracis* at about 150 g and *Monastrea* at about 1330 g – ten times heavier (but not so different in amount of space occupied). The percentage weight gain in *Acropora* and *Madracis* was thus consistently higher than in *Montastrea*, presumably because of their relatively greater surface areas. Table 7.2 lists the mean monthly percentage weight gains for these and some other experimental colonies and shows the importance of colony size in interpreting weight gain measurements. The differences in percentage weight gain for different sized colonies of the same species were probably the result of the effect of size on the surface area to volume ratio. Living surface area measurements are the obvious ones to have but are difficult to get because of the uneven nature of the surface of the skeleton. Bak (1976) measured living surface areas of his *Montastrea* and found no differences in calcification rate per square centimetre between large and small colonies.

Thus, in terms of surface area, or amount of living coral tissue, calcification rates within a species are independent of colony size – corals do not become senile. This conclusion is supported by linear

Table 7.2. *Approximate mean monthly percentage weight gains of various corals from Curaçao (growth of A. palmata measured at a depth of 3 m and of the other species at 12–15 m)*

Species	Growth form	Approximate colony size	Monthly % weight gain
Acropora palmata	thick branches	small (500g)	12
Acropora palmata	thick branches	large	7
Madracis mirabilis	thin branches	500 g	6
Agaricia agaricites	foliaceous plates	500 g	3
Montastrea annularis	massive heads	50 g	4
Montastrea annularis	massive heads	1500 g	1

Data from Bak (1976)

measurements of annual growth rings in corals. Growth rings have recently been demonstrated by the use of X-radiography on slices through coral colonies (Fig. 7.5), they result from seasonal changes in the density of the skeleton as it is laid down and provide a very useful retrospective way of studying coral growth (skeleton density also varies between colonies and between species adding to the difficulty of using linear growth as a measure of calcification.). In Fig. 7.5 one can look back several years and see that radial growth has not changed with size. Thus linear growth is steady and weight increments gradually increase with size; but what of the living surface? In branched colonies surface area probably increases exponentially with time, in proportion to colony weight, but in

Fig. 7.5. X-radiograph of a thin slice through the centre of a colony of the coral *Porites lutea*. Annual growth bands are visible; mean radial growth rate in this species is 7.6 mm per year. From Highsmith (1979).

massive colonies the growth of surface must lag behind the growth of volume (weight). This was pointed out by Barnes (1973) who suggested that as massive rounded corals increased in size, a critical radius would be reached at which tissue growth would be limited by the low rate of increase in skeleton surface. Any increase in size above the critical radius would require a change of form to accommodate tissue growth. Barnes suggested that plate-like skirts around the edges of massive heads, or irregularly lobate or knobbly surfaces, are growth strategies which function to increase surface areas in large massive colonies. The applicability of Barnes's model is unclear since one can find huge, regular, hemispherical colonies, especially of brain corals (Fig. 7.3), which appear long past any sensible critical radius; but the point that skeleton and tissue growth can be viewed as separate phenomena is an important one to make and I will return to it below when discussing growth forms.

7.3.1 The effects of light and depth

As explained in the introduction to this chapter, light promotes calcification. Laboratory experiments on a variety of species have shown that calcification proceeds 2–20 times more rapidly in the light than in the dark (Buddemeier & Kinzie, 1976). On the reef itself the position is complicated by various factors affecting light such as depth, turbidity and surrounding reef structures which may shade corals. In shallow water there is evidence that some species show reduced growth because the light is too bright. In general, however, growth responds positively to light. In Fig. 7.6 the mean monthly growth of *Montastrea annularis* colonies can be seen to follow closely the monthly variations in the number of hours of sunlight. These colonies were all living at the same depth under the same environmental conditions and growth was not found to be correlated with the small seasonal temperature fluctuations, nor with the seasonal change in daylight hours. The close correlation with sunlight hours of growth at this depth (15 m) shows the importance of overcast skies in reducing available light.

Larger differences in available light are, of course, correlated with depth and there have been several studies of the effect of depth on growth rate. The expected result is a considerable decrease in growth rate with depth, but this is by no means always

found. Fig. 7.7 shows the radial growth responses of two massive
species from Hawaii; in both cases there is a statistically significant
reduction in growth with depth but the variation between colonies
at any one depth may exceed the variation due to depth over the
range studied. Bak (1976) studied increase in weight in relation to
depth in a massive species, *Meandrina meandrites*, and a foliaceous
plate-like species, *Agaricia agaricites*. *Meandrina* colonies
weighing about 2 kg each were studied at 6 and 25 m and were
found to grow about 25% more slowly in the deeper water.
Agaricia colonies weighing about 500 g each, however, were found
to grow slightly faster at 24 m than at 13 m – the reverse of the
normal pattern. *Agaricia* is characteristic of deeper water, being
abundant on the fore-reef slope, and may be inhibited by high light
intensities. This response of *Agaricia* suggests the possibility that
species showing particular depth preferences may be adapted to
using light most effectively at the intensity and spectral composition characteristic of the preferred depth. In line with this it is
worth pointing out that *Acropora* spp., the most rapidly growing
corals, are almost always found in shallow water – the growth rates

Fig. 7.6. The relationship between monthly coral growth and
monthly hours of sunlight. (*a*) mean monthly percentage
increase in weight of about 12 colonies of *Montastrea annularis* at
15 m. (*b*) number of hours of sunlight per month during the same
period. After Bak (1974).

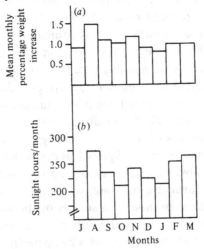

shown for *A. palmata* in Table 7.2 were measured in corals growing
in 3 m of water whereas the other species in Table 7.2 were growing
at 12–15 m. In conclusion, therefore, the relationship between light
and growth is certainly not a straightforward direct proportion.
Many species have depth ranges that span light intensity ranges of
over two orders of magnitude but show a much smaller range of
growth rates. Phenotypic adaptation of the calcification process
must be important here, as must also be the response of growth
form to depth.

7.4   Growth forms

An early contribution of divers to coral taxonomy was to
show, by extensive observation and collection, that many of the
distinctive colony shapes previously thought to represent different

Fig. 7.7. Mean annual linear growth rates of colonies of two coral
species with relation to depth. (*a*) *Favia pallida*; (*b*) *Porites lutea*.
Annual growth was measured as the distance between bands in
X-radiographs (see Fig. 7.5); each point is the mean for a single
colony. For discussion see text. From Highsmith (1979).

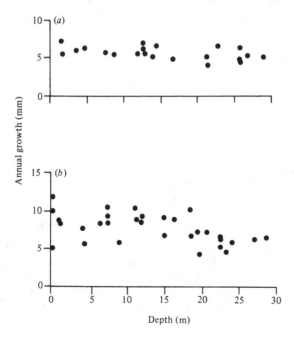

species, were actually growth forms of single species. Interme-
diates between the extreme shapes were found and the shapes
themselves were often seen to be associated with particular environmental conditions. Thus colony shape lost ground as a
taxonomic character and the number of species of corals was
reduced. Divers have long since made up the deficit by further
exploration, especially in otherwise inaccessible environments, but
the problems raised by the growth forms remain. Even in those
cases in which the association between form and environment is
clear, the factors which cause colonies to grow in characteristic
forms are often a matter of conjecture. The two environmental
variables which seem particularly involved in modifying colony
form are water movement and depth.

7.4.1   Effect of depth

A frequently noticed change of shape with depth,
described for a wide variety of species which form rounded heads in
shallow water, is to form progressively more flattened colonies as
the depth increases (Fig. 7.8). The flattened, plate-like colonies
found at depth are roughly horizontally aligned with the upper side
covered with living coral tissue. The underside, except for the

Fig. 7.8. The deeper-water plate-like growth form of the coral
Montastrea annularis; compare with Fig. 7.3c. Sediment has
accumulated in the centres of some of the plates. 20 m, Tobago,
Caribbean.

20 cm

periphery, is dead skeleton usually colonized by bryozoans, sponges, etc. Studies of growth in shallow and deep colonies of *Montastrea* (Fig. 7.9) show that rounded heads form by rapid upward radial growth grading to somewhat slower lateral growth around the sides of the colony. Plate-like colonies on the other hand, grow upwards very slowly but grow laterally more rapidly by adding skeleton around the edges. Graus & Macintyre (1976) were able to simulate these growth forms using a computer program in which the only variable was light. From published data they calcu-

Fig. 7.9. Outline drawings of sections through the centres of contrasting growth forms of *Montastrea annularis* which had been live-stained *in situ* with Alizarin Red – S one year before collection. The zig-zag line shows the staining horizon with the subsequent year's growth above it. The dotted lines indicate the growth directions of individual calices. (*a*) is the coral-head growth form characteristic of shallower water; (*b*) is the plate-like growth form characteristic of deeper water. Traced from a photograph in Dustan (1975).

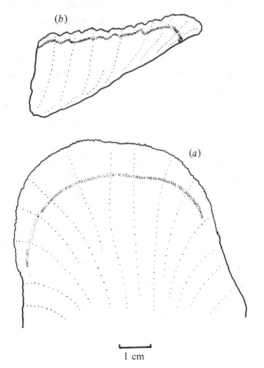

1 cm

lated the intensity of light from various directions at various depths, devised a proportional relationship between light and calcification, and assumed a minimum of 10% growth of living surface per year. Running the program they found that, in shallow water, the all-round light intensity produced sufficient radial growth to form a rounded coral head and to accommodate the assumed surface growth; in deeper water, however, upward growth was insufficient and the programed tissue growth therefore had to be supported by outward calcification around the edges, forming a plate-like colony. In addition they found that at intermediate depths the program predicted a columnar growth form characterized by rapid upward growth but slow lateral growth; such colonies are often observed at intermediate depths on the reef.

Graus and Macintyre's model bears some relation to that of Barnes (1973), mentioned above. In both cases a potential conflict between skeleton growth and tissue growth is resolved by an alteration of shape achieved by an uneven distribution of calcification. The growth rings in Figs. 7.5 and 7.9 show an uneven distribution of growth per square centimetre of living surface which, in rounded colonies, can be explained by the lower light intensity on the sides of colonies. In plate-like colonies, however, the edges are growing in the same light as the centre of the colony, but they grow faster. Thus the evidence implies the existence of a link between tissue growth and skeleton growth other than the obvious one that tissues produce skeleton which provides space for tissue growth. It looks as though tissues produce skeleton but tissue *growth* produces *more* skeleton, and that tissue growth in flat colonies occurs only round the periphery. If tissue growth promotes skeleton growth then it can be argued that the influence of light on skeleton growth is not primarily on calcification but rather on coral nutrition through photosynthesis by the zooxanthellae; carnivorous nutrition, round the edges of plate-like colonies, might also support skeleton growth. Whatever the answer to this puzzle, it is likely that in each coral species an underlying biological program exists and determines growth rates in different parts of the colony according to the environment to achieve the characteristic growth form. A clear and contrasting example of this is the growth of branching species where the major part of the linear growth of a branch results from growth at the tip – this region grows much faster than the proximal parts of the branch.

Evolutionary theory suggests the existence of a selective advantage associated with growth forms. An obvious advantage associated with the deep water, upward facing, plate-like form is that it harvests the maximum light per square centimetre of living surface; an important property in this dim environment. An attendant disadvantage is that flat colonies are liable to sedimentation: pools of sediment may cause partial mortality in the centres of such colonies (Fig. 7.8). However, a distinctive variety of the plate-like form, the flow-sheet, seems designed to shed sediment (Fig. 7.10). Flow-sheets occur on the sides of buttresses and cliffs in intermediate depths (10–20 m) and are orientated, like tiles on a roof, with the living surfaces facing obliquely upwards. The orientation

Fig. 7.10. A colony of the massive coral *Montastrea annularis* showing the flow-sheet growth form on its sloping sides. The colony is about 1 m across. Compare with Figs. 7.3c & 7.8. 10 m, Tobago, Caribbean.

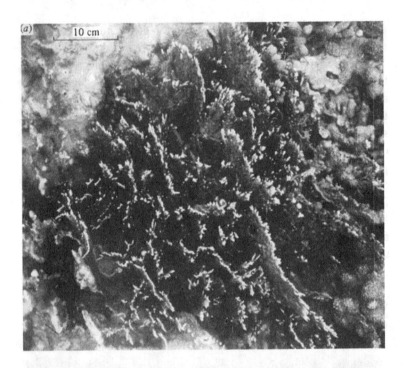

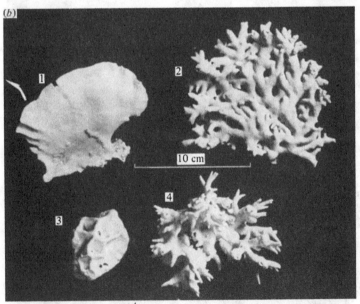

appears ideal both for harvesting light, which on a cliff is mostly coming in obliquely, and for shedding sediment.

7.4.2 Effect of water movement

The orientation of gorgonian and black coral fans at right angles to the prevailing current or surge has already been described (Chapter 2). The fan-like growth form combined with this orientation might seem suitable for stony corals since they are partial passive suspension feeders, but few species grow in this way. The reason is probably the brittle nature of the skeleton: sea fans bend when the rate of water movement increases, as occurs during storms, but fan-shaped stony corals would break. Further, the zooxanthellae of reef-building corals would not prosper on a rigid fan since most of the surface would face sideways rather than upwards (the zooxanthellae of symbiotic gorgonian sea fans often face upwards as the fans bend to and fro in the swell). Stony corals which do form fans at right angles to the water movement include the hydrozoan stinging coral *Millepora* and certain deeper water, non-symbiotic corals.

The various species of *Millepora* form a bewildering array of shapes (Fig. 7.11). These include an encrusting form which can grow over the skeleton of gorgonian fans, eventually completely replacing the living gorgonian tissue. But *Millepora* can also form its own fans which may be fenestrated or may grow as solid plates. *Millepora* fans develop in shallow water where there is wave action, they are quite small – 6–10 cm from base to edge – and grow at right angles to the surge, often in parallel rows from an encrusting base. Presumably their small size protects them from breakage, and, since they arise from an encrusting base, they can be replaced if they do get broken. A related but stronger growth form,

Fig. 7.11. Growth forms of the stinging coral *Millepora*. (*a*) view from above of a group of fenestrated fans at right angles to the prevailing wave action; 3 m, Tobago, Caribbean. (*b*) growth forms collected from various parts of the Caribbean: 1, side view of a fan-like plate; 2, side view of a fenestrated fan; 3, view from above of a reticulation of plates; 4, view from above of a bushy form. Photograph by the University of Reading Photographic Department.

probably adapted to turbulent conditions, consists of upward growing plates joined roughly at right angles to each other to form a reticulate pattern of plates arising from the encrusting base (Fig. 7.11).

The deeper water, non-symbiotic stony corals which form fans include the little hydrozoan coral *Stylaster*, sometimes found in cryptic environments on coral reefs. In deeper water in the open, exposed to ocean or tidal currents, *Stylaster* forms small lacy fans 6–8 cm high. I have observed them in the Caribbean and in the Indian Ocean. Much larger non-symbiotic stony fans are formed by the coral *Dendrophyllia*. These dark green or black fans may be half a metre across but live exposed to fairly gentle currents. They have been observed below 40 m exposed to tidal currents off the Aldabra reef, and I have seen them off Mauritius at the base of the reef at about 30 m.

A contrasting growth form which develops in response to directional wave action is shown by the elk-horn coral *Acropora palmata* in the surf zone on Caribbean reefs. In this environment the branches of this tree-like coral are rounded in transverse section

Fig. 7.12. View from above of the sheltered growth form of the elk-horn coral *Acropora palmata*; compare with Fig. 6.3. The colony on the left is about 1 m in diameter. 3 m, Tobago, Caribbean.

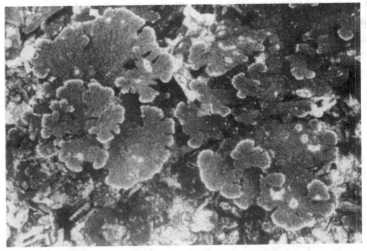

and predominantly orientated parallel to the surge. In quieter environments the branches are flattened, face upward, and show no preferred alignment (Fig. 7.12). The parallel orientation in surge is clearly adapted to minimize bending stresses while the rounded section adds strength to resist bending and streamlines the branch to the up and down movements often encountered in heavy surf. In contrast the flattened radial growth form is fragile but would function as a good light-harvesting surface.

8

Biological interactions with reef corals

8.1 Introduction

This chapter is concerned with the biological environment of reef corals. A coral's eye view of its organic neighbours would include spatial competitors threatening to overgrow it, predators threatening to eat it, boring organisms tunnelling into it, and a variety of animals sheltering under it and creeping and swimming over and amongst it. Coral reefs are particularly diverse environments so there are a large number of possible interactions.

One approach towards understanding the nature of reef ecology is to suggest that this diversity is the result of a long, stable evolutionary history during which specialization and co-evolution occurred. This viewpoint sees the reef as an ordered system with each organism tuned by evolution to fit a particular, narrow niche. The specialization, and consequent narrowness, of ecological niches allows the existence of more niches, hence diversity. However, not all reef organisms are restricted to reefs and many may be found living amongst boulders, or pier pilings, where there is little coral growth. This implies opportunism rather than specialization. An alternative approach, therefore is to grant that there are some co-evolved relationships – e.g. symbionts, commensals – and some specialization, but to suggest that the majority of species are to some extent opportunistic and that diversity arises partly through the complexity of reefs as a habitat (a high level of environmental heterogeneity leading to more niches) and partly through chance. This view of reefs is of a more chaotic than ordered environment about which a biologist would be unwise to make precise predictions. These two views are discussed by Cameron (1977) with relation to plagues of the coral-eating starfish *Acanthaster planci*: plagues which should not occur in an orderly, balanced system. The truth of the matter, however, seems to be that there is both order and chaos. Being close to the surface of the sea, chance and the weather are part of the nature of reefs and there will always be room for opportunistic species. Chance constantly supplies new habitats for colonization and chance governs the identity of colonists since not all larvae or juveniles are equally available at all times. On the other hand, potential colonists probably have preferences for particular ranges of physical and biological conditions and sort themselves out (or are sorted by the environment) in a general way into the various broadly defined reef habitats – reef flat, buttress zone, etc. – so that assemblages occur which are characteristic in general but unpredictable in detail. Thus the coral's eye view of its neighbours is of a series of possible interactions: some highly probable and some less so.

My intention in this chapter is to examine those interactions which directly affect the lives of corals and to consider the defences which the corals may use to avoid being disadvantaged.

8.2   Spatial competition

A hazard which must be faced by all sessile organisms is that of overgrowth (see § 3.3). This may take the form of a physical smothering or of a less direct deprivation of resources. Encrusting organisms may simply grow over their neighbours, killing them by depriving them of all access to the surrounding water. Erect, branching or plate-like organisms which overshadow their neighbours have less drastic effects, but by occupying space above them they may reduce the available food by filtering some of it out before it reaches the understorey, and they reduce available light for reef corals and algae. As was mentioned above (§ 7.2), branching corals may grow as a canopy above an understorey of slower growing, massive corals. One aspect of inter-coral competitition in this case is the effect on the understorey of shade cast by the canopy.

Shading depends on the density of the canopy. A canopy of the loosely branched Caribbean coral *Acropora cervicornis* reduces light intensity by only about 50%, but the more densely branched species of *Acropora* growing in the tabular form common in the Indo-Pacific reduce the light by up to 95%. Sheppard (1981) investigated coral communities growing in the shade of the latter at

10 m deep on the Great Barrier Reef but found little difference between shaded and unshaded assemblages. He pointed out that several coral species have wide depth distributions spanning more than two orders of magnitude of light intensity. Thus the understorey corals at 10 m, assuming shade adaptations, are probably still above their compensation light intensities and so are not limited by shading. Deeper on the reef, however, shading may be more important.

There is not much that a coral can do about being overshadowed by a faster growing, branching species, but if the two come into physical contact then the coral's biological defences are stimulated. Lang (1973) showed that corals placed in contact in the laboratory would extend mesenteric filaments from the contacting polyps. These filaments can digest and kill the tissues of the alien coral and Lang found that there was always a winning and a losing species, the tissues of the latter being destroyed by the mesenteric filaments of the former. By testing various combinations it was found that coral species formed a linear hierarchy with the most aggressive (most destructive mesenteric filaments) at the top. Members of the Mussidae, Meandrinidae and Favidae were usually fairly high up the hierarchy and the Poritidae were fairly low down. Lang suggested that aggressive mesenteric filaments were the means by which slower growing corals maintained themselves in competition with faster growing species.

A contrast to Lang's work is that of Sheppard (1979), carried out on the Chagos Atolls in the Indian Ocean. Sheppard recorded 683 naturally occurring interactions observed during dives at various depths on the reefs. Interactions could be recognized by close juxtaposition of species with a band of bare skeleton separating the living tissues. Winners and losers could be distinguished by the appearance of the opposing polyps and by the identity of the bare skeleton (that of the loser laid bare by the mesenteric filaments of the advancing winner). The hierarchy of aggression found by Sheppard differed from that of Lang in several ways. First, it was less clearly linear; instead of presenting a hierarchy, Sheppard grouped the species into three classes, aggressive, intermediate and subordinate. Second, there was little relationship between taxon and aggressive class: many families and even some genera contained both aggressive and subordinate species. Third, where there was a relation between taxon and aggressiveness it was usually different from that observed in the Caribbean; e.g. the Mussidae were highly aggressive in the Caribbean but intermediate on Chagos. An exception here was the genus *Porites*, subordinate in both areas. Lastly, Lang's suggestion that aggression is a means by which slower growing corals can compete with faster growing species was not supported by Sheppard's findings. The most aggressive Chagos corals included some fast growers such as *Acropora* spp. and also included several common species which formed extensive nearly monospecific zones. Aggression was therefore usually correlated with evident ecological success rather than being a defensive property of otherwise disadvantaged species. However, even on Chagos, aggression was not always correlated with success: extensive zones dominated by *Porites*, the most subordinate genus, also occurred.

Lang's laboratory study appeared to involve aggression by mesenteric filaments only, but Sheppard's observations were of naturally occurring interactions in which other mechanisms may have been operating. One such possible mechanism, described for Caribbean corals, involves extra long tentacles called sweepers which project from the polyps at night and wave about in a characteristic fashion. Sweepers can sting encroaching corals and other organisms using an especially large battery of nematocysts at the tentacle tip. Richardson *et al.* (1979) studied the sweeper tentacles of *Montastrea cavernosa* by diving at night on the Discovery Bay reef in Jamaica. They photographed sweepers more than ten times longer than normal tentacles apparently stinging a neighbouring sponge (Fig. 8.1). Experimentally presented colonies of *Montastrea annularis* were also attacked by sweepers. *M. annularis* is higher in Lang's aggressive hierarchy than *M. cavernosa* and would be expected to overgrow it; however the two species are often found side by side, usually separated by a space about 3 cm wide (Fig. 8.2). This is narrow enough for sweeper tentacles to cross, but too wide for mesenteric filaments. Thus the sweepers may prevent overgrowth by the more aggressive species. Richardson *et al.* put it this way; 'The long, waving sweeper tentacles may act like a boxer with his glove on the opponent's head, keeping the

Fig. 8.1. Close-up of polyps of *Montastrea cavernosa* at night showing sweeper tentacles adhering by enlarged stinging tips to the surface of a neighbouring sponge (top right). Photograph by P. Dustan. From Richardson *et al.* (1979).

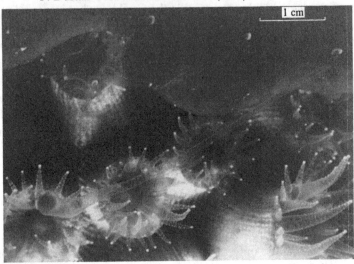

Fig. 8.2. Close juxtaposition of colonies of the massive corals *Montastrea annularis*, on the left with small polyps, and *M. cavernosa*, on the right with large polyps. Antagonistic interactions result in a band of bare reef-rock about 3 cm wide separating the species. See text for details. 12 m, Tobago, Caribbean.

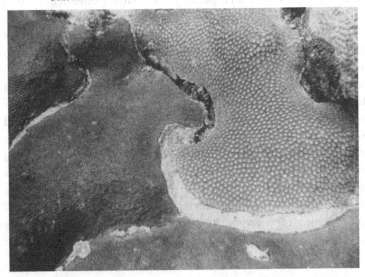

opponent's punching arms (mesenteric filaments) just out of reach of his vital organs.'

Organisms other than stony corals also pose an overgrowth threat, but less is known about them. Alcyonarian soft corals were found by Sheppard (1979) to be able to overgrow most scleractinian stony corals. The overgrown tissues are apparently killed by smothering rather than by any specific aggressive action, but clearly the defences of the scleractinians do not work against alcyonarians. In the Caribbean, colonial ascidians have been observed to overgrow corals which are high in Lang's inter-coral hierarchy: mesenteric filaments may not work against ascidians either. Colonial ascidians can move over coral surfaces by growth and regression or by a more rapid form of motility, timed in the laboratory at about 0.4 mm hr^{-1}. They smother the living coral at the leading edge, and at the trailing edge leave bare skeleton behind (Birkeland *et al.*, 1981).

A rather different type of spatial competition is provided by algae. These organisms grow much faster than corals and can rapidly smother small colonies. Coral death follows as a result of silt accumulating around the bases of the algae. The fact that newly settled corals survive at all in the face of this competition owes much to the herbivores of the reef. Grazing sea urchins and fish keep the algae under control and allow the small corals time to grow; eventually the corals out-compete the algae by growing too large to be smothered.

8.3 Predation on coral

Animals which graze live coral are liable to be stung by the nematocysts. It is perhaps not surprising, therefore, that relatively few of the familiar reef animals feed on coral. Common grazers such as urchins and parrot fish (Scaridae) mostly feed on algae and although they occasionally graze live coral, the bulk of their food is obtained from grazing the surfaces of reef-rock. Here they may graze newly settled corals along with the algae but, as mentioned above, most studies show that moderate grazing has a net beneficial effect on the survival of young corals. Heavy grazing pressure, however, is as bad as no grazing, and where it occurs young corals only survive in protected microhabitats such as crevices and overhangs where they are out of reach of the grazers.

A few fish feed more regularly and directly on corals. Those which rasp the living surface or bite off branch tips include the trigger fish (Balistidae), file fish (Monacanthidae), puffer fish (Tetradontidae) and some wrasse (Labridae). In contrast, some butterfly fish (Chaetodontidae) pluck polyps from the surface without damaging the skeleton. Invertebrate predators of coral include gastropods and worms. In the Caribbean the small (1 cm) coral snail *Coralliophila abbreviata* was found to eat 9 cm^2 of coral tissue in 24 hr and the large (15 cm) errant polychaete *Hermodice carunculata* ate 3 cm^2 in 3 hr. At normal population densities of 13 and 1 m^{-2} respectively, however, neither was considered likely to cause much damage on normal reefs. Following hurricane damage to the Discovery Bay reef in Jamaica, however, grazing by *Coralliophila* was observed to be seriously hindering regeneration. *Coralliophila* is not easily seen amongst the coral but the large brightly coloured *Hermodice*, protected by white bunches of irritant spicules along each side of its dorsal surface, roams over the corals in full view. I have seen this worm on the stinging coral *Millepora* with a branch of the coral, about the same diameter as the body of the worm, thrust deep into its pharynx.

Wounds caused by predators heal fairly quickly, the rate depending on the size of the wound. Bak *et al.* (1977) made experimental lesions on coral colonies to test wound-healing ability and found that algae and boring sponges soon settled on the exposed skeleton. Coral tissue, however, was able to overgrow these competitors and most small lesions of 1 cm^2 had healed in 80 days (Fig. 8.3). Larger lesions of 5 cm^2 took much longer to heal and some actually got bigger during the period of observation. Bak *et al.* also found that the rate of healing was influenced by the type of wound and by the species of coral. Lesions in *M. annularis* in which the skeleton had been damaged healed more rapidly than ones in which just the tissue had been removed, whereas *Agaricia agaricites* behaved in the opposite way and generally healed less rapidly than *M. annularis* (Fig. 8.3). Such differences might be expected to influence the competitive balance between species in the presence of predators. For instance, *M. annularis* should survive less well when preyed upon by butterfly fish than by trigger fish.

8.4   The crown-of-thorns starfish

The most carefully studied of all coral predators is *Acanthaster planci*, the crown-of-thorns. This large (30 cm diameter) Indo-Pacific starfish is normally rather rare on coral reefs, with densities in the region of 0.01 m^{-2}, but occasionally populations increase to 'plague' densities with hundreds of starfish concentrated into quite small areas. When starfish plagues develop the predation pressure on the corals is severe. For instance, in 1969 divers reported the destruction by starfish of 90% of living corals along 38 km of coastline in Guam from low water down to 65 m deep. Similar destruction has since been reported elsewhere in the Indo-Pacific region and the starfish have been studied especially in the Red Sea and on the Great Barrier Reef.

Fig. 8.3. Recovery rates of small (1 cm^2) lesions on the surfaces of colonies of two species of coral. For tissue lesions just the living tissue was removed but for tissue + skeleton lesions the underlying superficial skeletal elements were also scraped away. $n = 30$ in each case. Modified from Bak *et al.* (1977).

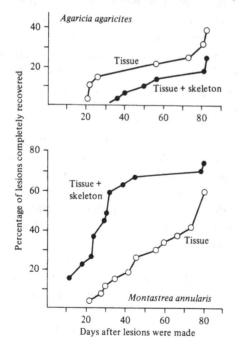

A. planci feeds by climbing onto the surface of the coral, everting its stomach and digesting the coral tissues. It does not harm the skeleton and leaves behind large bare white patches of skeleton which are quickly colonized by algae. It shows preferences between coral species which may be related to the nematocysts since some observations suggest that the tube feet may be stung as the starfish climbs up the coral. At normal population densities the preferences of *A. planci* may influence the composition of the coral community, but when the starfish are hungry most species of coral are consumed, hence the general destruction caused by plagues. There appears to be some tendency for normal populations of starfish to aggregate while feeding or spawning but the occurrence of plague densities has not so far been clearly explained.

Several theories are available (Endean, 1977) and fall under two headings: natural and man-induced. The natural theories take *A. planci* outbreaks as a dramatic effect of chance in the chaos view of coral reefs (see above, § 8.1). Sediments collected from the Great Barrier Reef containing *A. planci* spines have been used to suggest the previous occurrence of plagues (before the advent of industrial man) and the reasons suggested for occurrence vary from local destruction of reefs by storms (creating a hungry band of starfish which migrates *en masse* and forms the nucleus of a plague which grows as it advances along a reef), to cyclic or random changes in populations of starfish predators, or to random good years for starfish recruitment from the plankton. The apparently sudden recent appearance of starfish plagues is attributed in these theories to the rapid growth of diving: scientists had not previously been in a position to observe plagues. Man-induced reasons for plagues include undefined pollution effects aiding the survival of larvae or juveniles, and the documented reduction in the populations of certain predators of the adult starfish, especially the trumpet triton *Charonia* and the spiny puffer fish *Diodon* both of which are collected for sale as ornaments.

Whatever the cause of the plagues, the destruction of coral in affected areas is severe and recovery, especially for massive coral species is slow. Endean (1977) estimated that complete recovery might take 20–40 yr and added gloomily 'the possibility must be entertained that affected reefs will be reinvaded by *A. planci* during fragments rather than scripting, with builds (animations) on each slide to limit the amount of text seen at any one moment.

An issue that comes up in almost every talk I give is my vision. I can only see my slides if the podium on which my laptop rests is tall so that the screen is close. I can almost never read slides on nearby confidence monitors. And I can almost never see the laptop well enough to switch from the previous speaker's presentation to my presentation. So, I ask for help. I usually let the session chair know that she/he will have to cue up my presentation. I show up early to check out the podium and familiarize myself with any stairs or chords that may make getting to the podium more difficult. Most importantly, when it comes time for questions, I let the audience know that I will not be able to see they are trying to get my attention and ask the session chair to call on speakers for me. All this can be a little embarrassing, or sometimes quite embarrassing. The point is that we all have strengths and limitations as speakers. The key is to capitalize on our strengths and plan to minimize our weaknesses.

One of the strengths I have is experience, especially experience explaining complex ideas to non-experts. An essential part of my graduate training included teaching graduate-and undergraduate-level courses and advising other graduate students outside of my core discipline on the design and analysis of their research projects. In my first post-PhD position, I taught undergraduate Biostatistics to non-statistics majors for ten semesters. These experiences provided key foundations in communication that helped me to become more comfortable in a variety of communication situations.

You may not have as much opportunity as I have had to practice, but it is nevertheless important to take advantage of as many opportunities as you can.

4.4.2 Writing

Proofreading and revision are key to good writing, for even the most gifted writer. As a first attempt at practicing proofreading for scientific writing, consider Version 1 in the example used in Section 4.3.2, which is copied below for ease of reference. In Section 4.3.1, it was noted that this version contained details that were not necessary for convincing the audience the book *The Elements of Style* was a useful reference. Here, assume all the details are important for the intended purpose, but the passage still has too many adverbs and adjectives, and the complex punctuation leads to a lack of clarity. The first revision is improved by deleting unneeded adverbs and adjectives, and by breaking complex sentences into shorter sentences. The first revision is written in first person as is the original version and would be appropriate for casual settings or personal correspondence. The second revision is written in third person, which is often more appropriate for formal settings.

Original version:

> The most useful, relevant, and insightful instruction I have ever received on writing was via the excellent, well-known, timeless classic practical guide to writing by Strunk and White, The Elements of Style, the first vestiges of which appeared by way of Strunk's prodigious pen in 1918 and was published by Harcourt publishing in 1920, but was greatly expanded upon by E. B. White (Charlotte's Web) with the first edition of the expanded version of *The Elements of Style* published by Macmillan in 1959, and has since been named by Time magazine as one of the 100 best and most influential books written in English since 1923.

Revision 1:

> The most useful instruction I received on writing was via the classic book by Strunk and White, *The Elements*

of Style. The first edition of the book was authored by William Strunk in 1918 and was published by Harcourt in 1920. E. B. White expanded upon Strunk's version and the revised edition was first published by Macmillan in 1959. Time magazine named this version one of the 100 most influential books written in English since 1923.

Revision 2:

The Elements of Style is a classic book by Strunk and White that has helped countless authors improve their writing. Time magazine named it one of the 100 most influential books written in English since 1923. The first edition was authored by William Strunk in 1918 and was published by Harcourt in 1920. E. B. White expanded upon Strunk's version and the revised work was first published by Macmillan in 1959.

The following examples could be applied to either presentations or writing. The purpose of these examples is to illustrate the importance of using objective information to foster clear explanations.

The following example illustrates a philosophical approach to communication of stating the facts and letting the reader draw the conclusion for themselves, which is usually appropriate for a knowledgeable audience. The key feature is to minimize the use of adjectives. Philosophically, the goal is to show the reader, not tell them.

Example 1 – not preferred

The incidence of nausea among drug X-treated patients was 25.6%, which compares favorably with rates reported for other antidepressants (21%–37%).

Example 2 – preferred

The incidence of nausea among drug X-treated patients was 25.6%. Rates reported for other antidepressants have ranged from 21%–37%.

The following example illustrates the principle of using no more than one hedge word per sentence (compare Examples 1 and 2), but also shows how stating objective information removes the need for any hedge words and is more informative.

Example 1 – worst approach

Drug X has been generally safe and well tolerated by the majority of patients.

Example 2 – better approach

Drug X has been generally safe and well tolerated.

Example 3 – best approach

In randomized controlled trials, the incidence of adverse events among drug X-treated patients did not differ significantly from drug Y.

The following example illustrates the power of clear explanation in untangling the concepts of statistical significance and clinical relevance, which are easy for non-statisticians to confuse. In Example 1, a result comes from a large clinical trial in which a difference was statistically significant but small in magnitude and therefore unlikely to be meaningful in clinical practice. Example 2 comes from a small, pilot trial in which a large difference was observed but was not statistically significant.

Example 1

Although the mean cholesterol level for drug A was significantly lower than for drug B, the observed difference of 10 mg/dl is unlikely to yield an appreciable reduction in the risk of heart disease and is therefore of doubtful clinical relevance.

Example 2

The mean basal metabolic rates did not differ significantly between the exercise and no-exercise groups. However, the observed differences of 400 kcal/day would, all else equal, translate to a mean weight loss of 36 lb in one year. Therefore, a difference of this magnitude would be important if it was proven accurate by larger studies.

As with presentations, we each have differing sets of strengths and limitations in writing ability. Again, the key is to make the most of your strengths and minimize your weaknesses. If you are not at present a good writer, practice the principles in this chapter when writing emails or other short documents. Seek feedback. If you have a large writing project, such as a research paper, seek out a collaborator with strong writing skills, get their help, and learn from them.

Critical Thinking and Decisions under Uncertainty

Individual Factors

ABSTRACT

Chapter 5 is the final chapter in Part 1, working with self, and covers critical thinking and making decisions under uncertainty. The chapter begins by focusing on the ground-breaking work in Daniel Kahneman's book Thinking Fast and Slow. Kahneman introduces System 1 and System 2 as conceptual characters in our brains. System 1 is fast, intuitive, automatic, but prone to errors. System 2 is slow, deliberate, rational, but can be lazy. Each system has important advantages and limitations that are explained in detail. Understanding System 1 and System 2 thinking sets the stage for discussion of the many cognitive biases all of us are prone to. Extensive discussion is devoted to when intuition can

DOI: 10.1201/9781003334286-6

be trusted and when it can lead us astray. How quantitative analysis can be used to improve decision-making is also covered in detail. Chapter 5 concludes with practical ideas for putting these principles into practice, along with some real-world experiences from the author's career.

5.1 INTRODUCTION

As statisticians gain seniority and experience, we are increasingly called upon to make important decisions. Quantitative analysis is useful in making these decisions, but it is not the only thing we need. Consider the phase 3 go/no go decisions in the pharmaceutical industry, which are representative of many business applications. At the end of phase 2, companies decide if a drug should enter the next and most expensive stage of development, phase 3. Statisticians can estimate the probability the drug will succeed in phase 3 given a phase 2 result. The probability of success is a key factor, but it is not by itself a decision. The fundamental problem is to take the probability of success, which is an estimate on a continuous scale, along with estimates of other factors (such as cost, time of development, and other development opportunities), and distill all that information into a binary decision about whether development of the drug should be continued or terminated. How do we translate all the information into a decision?

Before digging into principles of decision-making, consider these two questions: (1) Is a decision an event or a process? And, (2) When faced with a key decision, what is the first thing we should do? Understanding the answers to these questions is foundational to many concepts that follow.

To begin answering these two foundational questions, consider the following five myths about how decisions are made that are dispelled in *The Art of Critical Decision Making* (Roberto, 2013).

- Myth #1: The chief executive decides. In reality: Strategic decision-making entails simultaneous activity at multiple levels of the organization. We cannot look only to the chief executive to understand why a company, organization, or school embarked on a particular course of action.

- Myth #2: Decisions are made in the room. In reality: Most decisions are made in one-on-one conversations or in small groups. Staff meetings often serve to ratify decisions that have already been made.

- Myth #3: Decisions are intellectual exercises. In reality: Decisions are complex social, emotional, and political processes. Social pressures for conformity and our inherent desire for belonging affect decision-making. Emotions can either motivate or paralyze us when making important decisions. Political behaviors such as coalition building, lobbying, and bargaining are important in organizational decision-making.

- Myth #4: Managers analyze and then decide. In reality: Strategic decisions often occur in a nonlinear fashion, with solutions arising before problems are fully identified and defined, or alternatives fully investigated. Decision-making processes rarely follow in a linear sequence, as classic models often suggest. In fact, sometimes solutions go in search of problems to solve.

- Myth #5: Managers decide and then act. In reality: Strategic decisions often evolve over time via an iterative process of choice and action. We often take some actions, make sense of those actions, and then make decisions about the next steps. Sometimes, managers decide and then do some analysis to confirm or justify the choice.

Dispelling these myths reveals that decisions are often first made at lower levels of the organization and then networked, confirmed, or funded at higher levels. But dispelling myths about how decisions are usually made is not the same thing as ensuring we are making good decisions. So, let's return to the two questions asked above.

As we will see later, good decisions are more likely if they result from a process rather than an event (Question 1). Good decision-making requires a series of interactions and events over time, within groups and across units of complex organizations (Roberto, 2013). The answer to Question 2 follows from the answer to Question 1. It is tempting when faced with a complex problem to immediately begin distilling the information. But that is a mistake. Instead, it is better to first consider what process should be used to make the decision and then follow that process rather than letting things take their own course (Roberto, 2013). Think of it like this: If you were beginning a complex and difficult journey, should your first act be to start moving or to plan the trip?

Making good decisions often entails a complex interplay between individual factors, group factors, and organizational factors (Roberto, 2013). At the individual level, understanding cognitive biases and how the mind works is key. Cognitive biases such as the sunk cost effect, loss aversion, or overconfidence can blind us to bad decisions and/or biased judgments. For example, in some situations, intuition is accurate and in other situations intuition cannot be trusted. By understanding what intuition is, we can learn when to trust and when not to trust our instincts (Roberto, 2013).

At the group level, understanding what it takes for teams to realize the promise of pooling the intellect, expertise, and perspectives of many people is paramount. Although diversity within the group holds the potential to enable better decisions than any individual could make, there is no guarantee. In fact, many teams make decisions that are inferior to those that the best individual within the group could make on her or his own (Roberto, 2013).

Moreover, decisions are not made in a vacuum. Environment shapes how we think, how we interact with others, and how we

the course of recolonization' and 'might be impoverished indefinitely'. Reefs in the process of recovery are not, however, dead. Algal production continues to support a diverse food web in which many reef species thrive. The species which are lost or reduced, apart from the corals themselves, are those which have evolved special relationships with corals.

8.5 Boring

A number of organisms bore into coral skeleton. These include algae, sponges, bivalves and polychaete worms. Boring takes place by acid secretion or mechanical abrasion or both. When the skeleton that is bored is the base of a living colony the effect is to limit the survival of the colony since boring weakens the base and makes collapse by gravity or wave action more likely. As was mentioned above (§ 6.5) only a small percentage (1–2% in shallow water) of the annual growth increment of a colony is removed by boring. But this loss is concentrated around the base because this is usually the only significant area of exposed skeleton. Thus as the colony grows the dead base not only becomes proportionately smaller, it becomes actually weaker as the small annual loss accumulates. A colony such as that shown in Fig. 8.4 would require only a slight push to turn it into a mobile boulder: corals with narrow bases are clearly more vulnerable than broad based colonies. Breakage at the base does not necessarily kill colonies, although it may do; it relocates them and can, by breaking them into pieces, lead to asexual reproduction.

Most animals which bore into coral skeletons are suspension feeders and Highsmith (1980) has shown that their abundance is linked to local ocean productivity. Highsmith looked at more than 700 coral heads in museum collections from all over the world and correlated the occurrence of boring organisms with published information on productivity. Fig 8.5 shows his results for boring bivalves; results for other groups showed similar trends and on a geographic basis led him to suggest a rank order from most boring to least boring as follows: eastern Pacific, western Atlantic, Indian Ocean, western Pacific. These differences lead to the interesting conclusion that east Pacific reefs should be more susceptible to storm damage than Indo-west Pacific reefs.

Fig. 8.4. A colony of the brain coral *Diploria* sp. on a narrow stalk (outlined and arrows). The fish beneath the coral is a goatfish and the urchins on the surrounding reef-rock are *Diadema antillarum*. The *Diploria* is about 25 cm across. 7 m, Jamaica.

Fig. 8.5. The relationship between the incidence of boring bivalve molluscs in coral heads and the productivity of the water where the corals were collected. From Highsmith (1980).

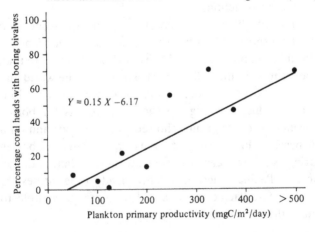

8.6   The effects of damsel fish

Damsel fish (Pomacentridae) are frequently herbivorous and sometimes planktivorous but do not generally graze coral. However, many species are territorial and may maintain the same small territory for months or years (§ 9.3.6). Their constant residence in and attention to their small areas of reef can have profound effects on the corals within territories. Damsel fish use their territories as a source of food, as a shelter, and for breeding purposes. They vigorously defend them against conspecifics and against other fish; they even behave aggressively towards divers. This defensive behaviour leads to less grazing inside territories than outside and the favourable conditions for algae lead to the development in the centres of territories of so-called algal lawns (Fig. 8.6).

Some species of damsel fish attack the living coral within their territories, exposing skeleton for algal colonization. In others however, the rich growth of algae, by itself, is sufficient to suppress the corals since it leads to sedimentation around coral bases and to death of basal tissues. In both cases more dead skeleton is exposed

Fig. 8.6. Photograph of a thicket of stag-horn coral *Acropora cervicornis* containing damsel fish territories the centres of which have become algal lawns – the darker circular areas each about 1 m across. Photograph by L. Kaufman. From Kaufman (1977).

Fig. 8.7. The reactions of small plankton-feeding coral fish (anthiids) to the approach of predators (a school of jacks, *Caranx*). (*a*) normal dispersion; (*b*) flight towards the reef; (*c*) hiding amongst the coral. The photographs, by G. W. Potts, were taken within a few seconds of each other. From Potts (1981).

to the attack of boring organisms and breakage and death of corals is greater within territories than outside.

Kaufman (1977) investigated damsel fish territories in Jamaica and his observations indicated that the fish can influence the species composition of the coral community. Kaufman suggested that the branching coral *Acropora cervicornis* is less badly affected in territories than the massive *M. annularis* because it can grow faster and redirect its growth out of the territories. Further, if boring causes it to break, the open branching structure should keep it clear of the substrate so that it can continue growing. *M. annularis*, in contrast, cannot rapidly redirect its growth and its shape would not keep it off the substrate if it fell.

8.7   Coral commensals

A large number of animals use coral colonies as a permanent habitat or a temporary shelter (Patton, 1976). Coral fish, especially the plankton-feeding *Chromis* spp., dart amongst branching corals when a diver or predatory fish approaches too closely (Fig. 8.7) and several fish species shelter amongst corals at night. More permanent tenants comprise a wealth of crustaceans and a few echinoderms which live amongst the branches and in crevices between coral heads. An idea of the numbers involved in

(c)

branching corals is given by Table 8.1; branching corals offer more protected space than massive corals but large numbers of cryptic animals – gastropods, polychaetes, brittle stars – may sometimes be found sheltering under massive coral boulders.

Many commensals, such as brittle stars, crinoids and porcelain crabs, are facultative associates which feed by filtering plankton and detritus from the water. Others however, especially some of the shrimps and crabs which live amongst coral branches, are obligate associates often with host-specific relationships. A common feature of obligate commensals is for sexually mature individuals to occur in mated pairs. Pairs are territorial and in any one species only one pair is normally found within a single coral colony. Different species, however, share colonies and partition resources by niche specialization.

Facultative associates probably have little effect on their coral hosts other than the inconvenience of being clambered over. Many of the obligate associates, however, feed on coral mucus and may, therefore, act as a drain on the coral's resources. Corals produce mucus to cleanse themselves of deposited particles, but, as noted above (§ 7.2) they may ingest particle-laden mucus as an extra

Table 8.1. *The effect of host size on the abundance of coral commensals.*
The host is the densely branched Indo-Pacific coral Pocillopora damicornis *which forms hemispherical to spherical colonies in shallow water. Relative ecospace is a measure of the space available for the commensals amongst the branches of the host colony. Species include shrimps, crabs and fish*

	Colony size			
	Small	Medium	Large	Very large
Number of colonies examined	6	18	10	4
Approximate colony diameter (cm)	12	20	30	40
Relative ecospace	1	3	8	20
Mean number of individuals per colony	7.7	21	33	65
Mean number of species per colony	4.5	7.5	7.5	8.3

Data from Patton (1974)

source of food. Mucus-feeding commensals deprive corals of this food. The small brightly coloured Indo-Pacific crab *Trapezia* has special combs on the tips of its legs with which it scrapes mucus from the surface of its host; if the host is not secreting mucus *Trapezia* pokes and scratches the polyps to stimulate secretion.

More obvious effects of commensals, in this case on the growth of the host, are produced by coral gall crabs. These tiny crabs are greatly modified for a sedentary existence with a huge abdomen developed in the female to hold an egg mass about as big as the body (Fig. 8.8). Female gall crabs of the genus *Hapalocarcinus* settle in the axils of young coral branches and their activities influence the growth of the coral so that the two branches form flat sheets which grow up and around the crab entombing it in a purse-shaped gall (Fig. 8.8). The edges of the flat coral sheets finally join leaving only a row of holes through which the crab can draw its respiratory stream and its food; gall crabs are thought to be suspension feeders but may supplement this diet with coral mucus. The related genera

Fig. 8.8. Coral gall crabs. (*a*) ventral view of a female *Hapalocarcinus* from the Indo-Pacific, showing the relatively huge abdomen; (*b*) galls in the coral *Pocillopora* formed by *Hapalocarcinus*, the topmost gall is the youngest and its edges have not yet closed; (*c*) side view of a female *Pseudocryptochirus* in its tunnel on the surface of the plate-like Caribbean coral *Agaricia*. (*a*) (*b*) from Potts (1915), (*c*) from Shaw & Hopkins (1977).

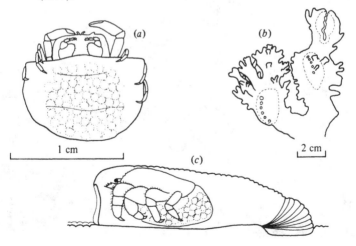

Cryptochirus and *Pseudocryptochirus* live in cylindrical caves or pits in the coral which grows up around them (Fig. 8.8). Male gall crabs are smaller than females and may share their galls.

Certain other animals may also be found embedded in the living surface of corals. These include some bivalves, polychaetes (Fig. 8.9) and barnacles which inhabit crypts in the skeleton underlying live coral tissue. In these cases it appears that the colonizing larva has settled on the living surface and that the latter has grown up around it, entombing it in skeleton but allowing the syphons, tentacles or cirri access to the outside water. This is advantageous for the commensal since the entrance to its crypt is protected by coral tissue. The pits and caves of gall crabs and the crypts of bivalves and worms may sometimes be taken over by small fish, especially species of goby several of which are obligate commensals. These fish live on the surface of their coral hosts and retreat into their pits when danger threatens.

Obligate coral commensals such as *Trapezia*, in view of their

Fig. 8.9. Christmas-tree worms *Spirobranchus* (Serpulidae) with crowns extended above the surface of a colony of the brain coral *Diploria* in which the calcareous tubes of the worms are embedded. The tentacular crowns are of several different colours and each is about 3 cm high. 8 m, Tobago.

feeding habits, are sometimes referred to as parasites. However, they have been observed to perform a protective service for their hosts against *A. planci* by pinching the attacking starfish and deterring it from feeding. It is a matter of conjecture whether the balance of benefit is in favour of the commensals or their hosts.

8.8 Coral diversity

It seems appropriate to end this chapter with a look at coral diversity in the light of the various facets of coral life discussed above and in the previous two chapters. Studies on coral dispersion patterns have often shown distinct aggregation – a patchwork of species scattered over the reef. In some species, such as *Acropora palmata* in the Caribbean, the patch is continuous along the coast at a particular depth and forms a zone, but frequently the patches are discontinuous and not limited to a particular depth. Again, in some of these cases, patches of a given species may appear to occur in special habitats, for instance on vertical faces; but this is not always found, and there are many other species which do not occur in patches at all but follow a more or less random distribution.

These patterns reflect both order, in the form of zones and patches in particular habitats, and chaos in the form of the many exceptions; but a diverse mixture is in any case expected on the basis of the wide array of conflicting pressures on the corals. Considering growth rate differences, aggressive competitive differences of various types with different effectivenesses according to the opponent, the chances of predation and the various recovery abilities according to the predator, the chances of interference by territorial fish, boring organisms and other commensals, and, one hardly needs to add, the possibility of storms, it can be seen that high diversity is almost inevitable. This is, in fact, a good broad illustration of the principle of the competitive network discussed above (§ 3.3) in which diversity is maintained by no single species having an overwhelming advantage over the rest.

In physically stressed environments the possibility of overwhelming advantage increases and this is no doubt why there is usually reduced diversity, and sometimes a characteristic dominant coral, in the surf zone. Removed from the 'protection' of their stressful habitat, however, such specialized corals are often at a startling disadvantage. Neudecker (1977) transplanted colonies of *Pocillopora damicornis*, a common branched shallow-water Indo-Pacific coral, from pier pilings at 1.5 m deep to the reef margin at 2.4 m and to the seaward slope at 15 and 30 m where this species does not occur, and compared their survival and growth with and without protective cages. Despite the lower light intensity, caged seaward slope transplants grew faster than reef margin transplants, those at 15 m growing fastest. But uncaged seaward slope transplants during the 7–9 days of the experiment lost about 25% of their mass to grazing fish (mostly trigger and butterfly fish) whereas the uncaged reef margin colonies were unaffected. Thus although *P. damicornis* does better in one respect in deeper water (growth) its failure in another (resistance to predation) is decisive. Perhaps the more rapid water movement and high light intensity of its normal environment protects *P. damicornis* from fish predators, but a further twist to the story is that it may be protected by territorial damsel fish which are relatively common in shallow water. Wellington (1982) showed that on balance the activities of damsel fish give a competitive edge to *Pocillopora* within territories: first, the territorial fish chase coral predators away, and second, they find difficulty establishing their algal lawns on *Pocillopora*, possibly because of this coral's tight branching structure, and so prefer to cultivate algae on other species.

9

Reef fauna: fish

9.1 Introduction

The diverse life of the reef, discussed previously with reference to its effects on the corals, has been studied from a variety of other angles. Some of these studies were referred to in chapters 2 and 3. For instance much of the work on orientation to water movements of sea-fans and crinoids (§2.2) was carried out by diving on coral reefs. And Jackson's (1977, 1979) ideas on the interactions of sessile organisms (§3.3) originated from studies on the cryptic fauna growing beneath plate-like corals. However, I have said little so far about fish.

One of the first things that divers say on emerging from their first dive or snorkel on a coral reef is 'What wonderful fish! Did you see that one with the . . .?' Coral reef fish are numerous and colourful; although in a given reef area there may be between 500 and 2000 species present, it is reasonably easy to recognize the common ones by sight and the clear warm water makes it possible for divers to watch at their leisure. There are many identification guides to reef fish, including some plastic ones that can be used underwater. Even among non-biologists interest is not directed only at the spectacular reef fish such as sharks and angel fish; the ordinary fish such as parrot fish, surgeon fish, damsel fish and butterfly fish get their share of attention because of their bright colours and amusing behaviour. Diving biologists are by no means immune to the attractions of reef fish; there is a considerable literature on their biology and behaviour, much of which has been reviewed by Sale (1980).

This chapter is not exclusively about fish, other animals are introduced where appropriate, but it is mostly about fish. It is also mostly about behaviour. The topics I have selected – feeding, social systems, symbiosis – are those on which most work has been done but I have not attempted to review this work, my intention is merely to point out some of the fascinating things about reef fish which have been revealed by diving studies. In the last section of the chapter, however, I return to the theme of Chapter 8 and look at fish diversity, how the space on the reef is utilized and the question whether the fish community is an ordered or a chaotic system.

9.2 Feeding

Reef fish can be separated into a number of feeding categories such as herbivore, planktivore, omnivore, carnivore, grazer, invertebrate feeder, piscivore, etc. Few of these however, are mutually exclusive. Thus grazers are mostly herbivorous but may graze live coral and may occasionally take small benthic invertebrates; carnivores often prey on both invertebrates and small fish. Surprisingly, most studies show a preponderance of carnivores in reef fish communities. Goldman & Talbot (1976) gave the relative proportions by biomass of planktivores, grazers, benthic invertebrate feeders and piscivores in their study area as 10%, 18%, 18% and 54% respectively. Presumably this situation is able to exist because of the presence of large but cryptic populations of herbivorous and detritivorous invertebrates such as echinoderms, crustaceans, molluscs and annelids. Despite the relative paucity of herbivorous fish their grazing activities are of considerable ecological importance. One study showed that fish graze about 60% of algal production, but nearly 90% of this is subsequently defaecated as detritus, providing plenty of food for detritivores such as small crustaceans and brittle stars.

Planktivorous fish include the little coral fish of the genus *Chromis* (Pomacentridae) and *Anthias* (Anthiidae) which hover in swarms 1–3 m above the coral catching individual zooplankters as they drift past (Fig. 8.7). Parrot fish (Scaridae) and surgeon fish (Acanthuridae) are common grazing herbivores, they often band together into schools composed of two or more species and forage over the reef. This schooling behaviour may offer protection from predators (see below) but is also advantageous when foraging in areas where territorial damsel fish (Pomacentridae) are common. Damsel fish territories offer the best grazing (§ 8.6) and the territory holder can be overwhelmed by a whole gang of raiders. Benthic invertebrate feeders may show specializations of structure and behaviour to deal with particular types of prey. Long snouted butterfly fish (Chaetodontidae) use their snouts to reach in amongst coral branches for shrimps and crabs. Trigger fish (Balistidae) use their strong jaws to deal with crabs, urchins and coral branches. Fricke (1973) described how a Red Sea trigger fish was able to attack and eat the long spined urchin *Diadema* by blowing it over and then biting the unprotected oral region. *Diadema* takes refuge from predators by being active mostly at night. By day it hides amongst the coral or, if in the open, aggregates into groups in which the long spines of neighbours stabilize individuals and protect them from attack.

Fig. 9.1. Some common types of coral reef fish; (*a*) parrot fish (15–130 cm), (*b*) butterfly fish (5–20 cm), (*c*) surgeon fish (15–30 cm), (*d*) damsel fish (5–15 cm), (*e*) trigger fish (20–50 cm). See also Figs. 8.4, 8.7, 9.2, 9.3, 9.7, 9.12.

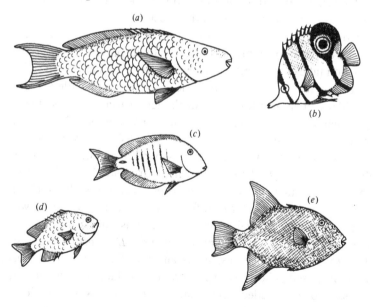

9.2.1 *Interspecific feeding associations*

Several predatory reef fish use behavioural tricks involving other species to facilitate their feeding. Ormond (1980), working in the Red Sea, described a variety of these interactions in which the predators were small to medium sized wrasse (Labridae) and sea bass (Serranidae). The simplest trick is for the predator to associate with a group of foraging herbivores; the foraging activity may disturb small invertebrates which the predator can seize as they scuttle away. Individual large foraging predators such as trigger fish and wrasse may also be followed for the sake of the small prey they disturb (see Fig. 9.6 for this behaviour in jacks). A less casual but similar association observed by Ormond (1980) and called by him 'interspecific joint hunting' involved a pair of similar sized predators of different species apparently cooperating to flush prey from coral heads. The advantage of this behaviour is assumed to be that each fish catches not just its own prey but also some of those disturbed by its partner. Ormond described associations between goatfish (Mullidae) and wrasse, both feeding on small invertebrates, and found joint hunting to be very common – 30–70% of the time, depending on the species.

A different type of interaction is for the predator to use another species as cover to enable it to creep up on its prey. This type of interaction takes two forms, riding and aggressive mimicry. In riding the trick is to swim so close alongside a larger non-hostile fish that the rider is mistaken for a part of the larger fish and so is able to approach its prey unobserved. The behaviour is called riding because it was first observed in the trumpet fish (Aulostomidae), a long thin piscivore with a trumpet-shaped mouth, which rides aligned along the backs of parrot fish. Ormond (1980) described riding in the small serranid *Diploprion*. This fish is about 12 cm long and feeds on smaller fish, fish larvae and crustaceans. It was observed to ride several species of larger fish including grouper (Serranidae), surgeon fish and parrot fish. *Diploprion* swims so close to its host that it resembles an extra fin; the species most commonly used as hosts were predominantly blue, like *Diploprion*, and so provided added camouflage. *Diploprion* was sometimes observed to break suddenly away from its host, apparently to secure prey. Ormond observed individuals of *Diploprion* for a total of 202 minutes and noted 32 bouts of riding with a mean bout duration of 15 seconds. Thus although the time spent riding was only 4% of total time, the frequent occurrence, about once every 6 minutes, indicates that it may be fairly important.

Aggressive mimicry is a trick in which the predator mimics a harmless species and associates with it. Prey apparently ignore non-predatory fish so the mimic is able to sneak up unobserved. Ormond (1980) watched this behaviour in two species of wrasse, one small and one medium sized. The small wrasse, a species of *Cheilinus*, has considerable ability to change colour to mimic other species. It was observed to mimic blue-black damsel fish, pale grey damsel fish, and goatfish which were sandy grey with a longitudinal dark stripe. Nine percent of its time was spent mimicing this species of goatfish. *Cheilinus* is also a rider and uses its ability to change colour (anything from pale grey through mottled orange and green to purple) to match its host. The larger aggressive mimic, the sling-jaw wrasse *Epibulus*, spent 11% of its time mimicking other species. *Epibulus* has some ability to change colour and can darken to mimic dark coloured surgeon fish or damsel fish (Fig. 9.2).

Fig. 9.2. Mimicry of the herbivorous surgeon fish *Zebrosoma* by the predatory sling-jaw wrasse *Epibulus* in the Red Sea.
Epibulus is the slightly larger fish in the centre, with elongated pelvic fins; it has darkened in colour to match the *Zebrosoma* and is approaching the coral to hunt for small fish. Photograph by R. F. G. Ormond. From Ormond (1980).

Ormond found that on average it would mimic other species once every 9 minutes for bouts of about a minute in length. Sometimes its mimicry was used to enable it to join surgeon fish raids on damsel fish territories and the possibility of it snapping up the defending damsel fish cannot be ruled out.

The existence of riding and aggressive mimicry gives some information about the abilities of prey species to detect potential predators. Evidently the size of the approaching fish is not a major factor since small predators can approach small prey under cover of a larger fish. Presumably the larger fish is not seen as a larger threat by the small prey because it expects to be ignored by a large predator (see Fig. 9.4). This argument does not hold, however, in the case of aggressive mimicry where the model is the same size as the mimic. Here one must presume that the prey recognizes the surrounding fish as belonging to harmless species (either herbivores or predators on a different type of prey) thus leaving itself open to attack by the mimic. All this suggests that reef fish are able to recognize each other at least at the species level, a suggestion further supported by the observations on joint hunting in which certain species combinations are much commoner than others.

9.2.2 *Predator–prey interactions*

Some rather different studies on predatory fish have been carried out by Potts (1970, 1980, 1981) in Aldabra. First Potts (1970) watched schools of small snappers (Lutianidae) sheltering in a reef channel, and observed their behaviour when attacked by groupers, large snappers and jacks (Carangidae). He observed that groupers and snappers approach the prey slowly and the school responds first by closing ranks and swimming away. Then those fish towards the edges of the school cascade out to either side and stream back past the predator keeping a set distance from it. On reaching the back, the school reforms and follows the predator for a short distance (Fig. 9.3). Attacks by the faster swimming jacks elicit the same behaviour but it all happens much more rapidly. I have observed similar behaviour in a large school of clupeids (herring-like fish) near rocks in shallow water on the Arabian coast. The school was under frequent but intermittent attack from the open water by jacks and from the rocks by snappers and groupers. Small groups of jacks would cruise in at high speed and the silvery school would cascade around them and move slightly closer to the rocks, from the shelter of which a snapper or grouper, manoeuvering slowly beneath, would suddenly lunge upwards producing a silvery fountain of fish. Schooling in prey species, and cascading behaviour when attacked, probably makes it difficult for the predator to single out a particular individual for capture: it is a strategy for confusing the attacker. The final following phase of cascading behaviour can also be used to dismiss the predator. I once observed a large school of clupeids cascading around a big barracuda (Sphyraenidae) in the Caribbean, but instead of merely following at a respectful distance many of the cascading fish were diving in to butt or bite the flanks of the predator. The barracuda clearly disliked this mobbing and left the area hurriedly.

In his later work Potts (1980, 1981) concentrated on the jacks and observed their predatory behaviour and the behaviour of their prey in shallow reef channels and in deeper water on the fore-reef. He found that jacks either hunt singly or in small groups. Groups may be composed of more than one species and more than one size

Fig. 9.3. Diagram illustrating the cascading response of a school of prey fish to the approach or attack of a predatory fish, in this case a snapper. From Potts (1970).

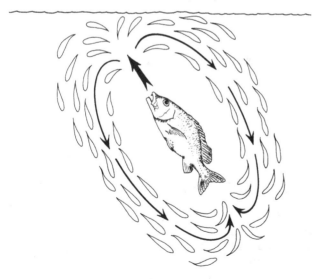

class. It is possible that by attacking in groups the cascading behaviour of the prey can be disrupted to the mutual advantage of the group members; however, groups produce more avoidance response than do single predators which may cancel out this advantage. Potts observed that the reactions of the prey depend upon their size and the size of the predator. In the channels potential prey included schools of small snappers 20–25 cm long and a variety of smaller fish, mostly wrasse and damsel fish 2–6 cm long living amongst coral rubble on the channel floor. The percentage of these two classes of prey which showed responses to attacks by various sized jacks were estimated and the results are shown in Fig. 9.4.

Fig. 9.4. Relationship between predator (jacks, *Caranx*) size and prey response in two size classes of prey in reef channels in Aldabra. (*a*) small prey 2–6 cm long living amongst rubble; (*b*) larger prey 20–25 cm long occurring in schools. From Potts (1980).

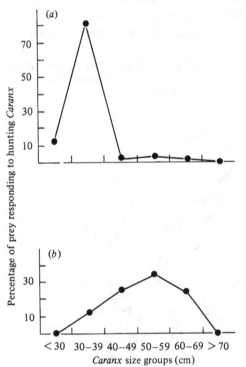

Clearly the rubble dwellers expect to be ignored by large predators whereas the snappers do not expect to be harmed by the smaller jacks. Potts found that the jacks used the cover of boulders to approach their prey and appeared to know where a school of prey would be. He noticed that hunting jacks would speed up while out of sight of their prey and would burst unexpectedly from around a corner straight into the school (Fig. 9.5). Potts' observations enabled him to construct a model of predation by jacks which is reproduced in Fig. 9.6.

9.3   Social systems

Populations of reef fish display a variety of social structures. They include species spread widely as isolated individuals, such as groupers, those with broadly overlapping home ranges, such as many of the wrasse, and those in which the individuals are tied to small defended areas or territories, such as some damsel

Fig. 9.5. Diagram of the attack route of a 60 cm jack (*Caranx*) on a school of small snappers in a reef channel in Aldabra. The arrows show the path taken by the jack and the longer the lines, the faster the fish was swimming. The open circle shows where an isolated snapper was driven into shallow water and attacked. After the attack the jack joined another jack which was swimming past. The irregular shapes show the positions of rocks and the stippled areas show where rocks overhung the water. After Potts (1980).

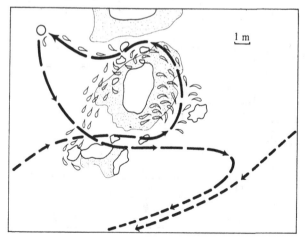

fish. They include species living in schools or groups either ranging over the reef, such as some parrot fish and surgeon fish, or remaining more or less stationary near particular coral formations, such as other damsel fish. They include monogamous pairs ranging widely, such as butterfly fish, or restricted to isolated habitats, such as the

Fig. 9.6. Diagram illustrating the predatory tactics of jacks, and predator–prey interactions. For further details see text. From Potts (1980).

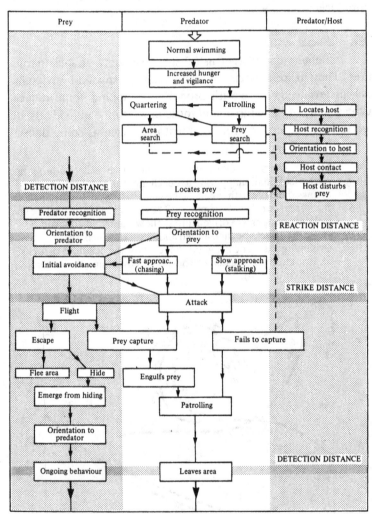

symbiotic clownfish (Pomacentridae). And they include species which adopt more than one social structure according to the time of day or stage in the life span of the individual. These social systems can be viewed as adaptations to a variety of different ecological roles. In particular they accord with the feeding and reproductive strategies and they aid in predator avoidance.

9.3.1   Interspecific groupings

One of the simplest cases is the mixed species school, several examples of which have been described. Ogden & Ehrlich (1977) observed large schools of up to several thousand individuals of two species of juvenile grunts (Pomadasyidae) on patch reefs in the Virgin Islands. These schools collected each day in traditional areas of the reef where there were large formations of branching corals. The schools are inactive during the day but at dusk the fish leave the reef by a number of traditional routes and swim out into surrounding sea grass beds where they disperse and feed. At dawn they return by the same routes and regroup at the schooling areas. The schooling habit presumably serves to protect the fish from predation and the schooling areas provide cover in the form of branching corals. Dispersal for night-time foraging suggests that the important predators are visually orientating fish such as groupers and jacks. The mixed species nature of the schools is associated with the juvenile status of the fish. When individuals reach a length of about 15 cm they leave the schooling areas: one species moves to deeper water while the other roams the patch reefs in smaller single-species schools.

Another example of mixed species schooling was studied by Itzkowitz (1974) in Jamaica. These schools were composed of 10–20, but occasionally up to 100, fish which grazed and foraged together during the day. Each school had a majority or core species, often ocean surgeon fish or parrot fish, which made up about 80% of the school. Associated species included various goatfish, wrasse, other parrot fish and other surgeon fish. These schools were loosely organized bands which tended to break up and reform as they moved amongst the coral heads. Small schools and isolated individuals would join up with larger schools as these swam past, but large schools tended to fragment as different parts of the school took different paths through the coral. Itzkowitz noticed that in any one school all members tended to be about the same size and he suggested that this was related to the swimming speed of the group – smaller fish would have difficulty keeping up. The advantages of grazing and foraging in groups were referred to above; they include protection from predation, the possibility of raiding damsel fish territories, and the disturbance providing food for predatory group members. Protection from predation is probably a fairly important function of the group: isolated individuals of group-forming species were observed to stay close to the coral but would join a group if one swam by and, in company, would swim freely in the more open areas of the reef.

9.3.2   Intraspecific groupings

Single species schooling is a common phenomenon amongst fish in general. On coral reefs pelagic visitors such as clupeids, tuna and mackerel usually occur in schools and some of the open water but near shore predators such as jacks and barracudas may occur in small groups or large schools. Even the mixed species schools referred to above are mostly one species. Association with conspecifics presumably carries most of the advantages noted above and also provides plentiful opportunities for social interactions, especially those concerned with reproduction. It is difficult to study the social structure and reproductive behaviour of fish which school in open water, whether they are pelagic or near shore, since they are free ranging and cannot be followed for long. Reef fish, however, often form single species groups which stay more or less in one area. Thus the same group can be watched day after day, and sometimes year after year, and the diver can learn to recognize some of the fish individually. This makes possible the experimental manipulation of groups to test hypotheses on social structure. Extended observations and experiments have revealed a variety of social systems amongst reef fish involving changes in behaviour, colour and sex during the life span of the individual.

9.3.3   Monandric social groups

A relatively simple example is provided by the small Indo-Pacific cleaner wrasse *Labroides dimidiatus* (Potts, 1973;

Robertson & Choat, 1974; see § 9.4.1). This fish grows to about 9 cm long and gradually changes colour as it grows. Juveniles are mostly black with thin dorsolateral and ventrolateral blue stripes. As the fish gets older the black area decreases to a black band along the side (Fig. 9.7). Females and males do not differ in colour. Individuals inhabit overlapping ranges which increase in area with the size of the fish from a few square metres in juveniles to 20–30 m^2 in adults. Areas (territories) towards the centres of these ranges are defended against similar sized and smaller conspecifics by means of aggressive lateral displays with fins erect and by attacks involving ramming and probably biting. Large adults usually occur in male–female pairs and jointly defend a territory within a large range which often overlaps the territories of several smaller adults and juveniles. These smaller fish all turn out to be female and the social system within each large adult range is therefore an hierarchical harem consisting of a single dominant male, a chief female, two to five subordinate but mature females and a few juveniles. Each subordinate female and juvenile has her own territory within the range of the dominant pair and the male makes frequent excursions about his range visiting the females and interacting both aggressively and sexually with them. Since all the small individuals in the population are females it is clear that the species is protogynous and males arise by sex reversal. Experimental removal of a dominant male confirms this: within a few hours the original chief female starts adopting characteristic male aggressive behaviours and within a few days is courting the other females. After 14–18 days sperm release is possible and sex reversal is complete. Thus it appears that sex reversal in large females is normally inhibited by the aggressive dominance of the resident male.

Fig. 9.7. The Indo-Pacific cleaner wrasse *Labroides dimidiatus*. (*a*) juvenile 2–3 cm long; (*b*) adult 6–10 cm long. Dark areas are black and pale areas are blue. From Potts (1973).

(*a*) (*b*)

A more complicated situation obtains in the small Indo-Pacific planktivorous coral fish *Anthias squamipinnis* (Anthiidae) (Shapiro, 1977, 1981). This species occurs in stationary groups of up to a few hundred fish hovering alongside vertical reef features. A commonly observed group size is about 30 fish and the space occupied by a dispersed group is about 1 m^3 per fish. Like *Labroides*, *Anthias* is protogynous, but unlike *Labroides* there is often more than one male per group. The mean sex ratio was found to be about eight females and several juveniles to one male, and the males were normally the largest individuals in the group. The experimental removal of one male from either single-male or multi-male groups was found to result in the rapid sex reversal of one only of the larger females. Since in multi-male groups sex reversal occurs in the presence of males it can no longer be suggested that females are inhibited from changing sex simply by male dominance (Shapiro, 1981).

9.3.4   *Diandric social groups*
Even more complicated arrangements are found in certain wrasse and parrot fish in which the majority of the population is protogynous but a sizable minority (e.g. 30%) starts out male and remains male throughout life. The latter are referred to as primary males whereas those which arise by sex reversal are called secondary males. At least two colour patterns occur in the lives of these individuals: juveniles and small but sexually mature fish are often referred to as 'drab' in contrast to the large primary and secondary males which are described as 'gaudy'. Sex reversal occurs within a fairly narrow size range and females turn directly into gaudy secondary males; primary males change from drab to gaudy at about the same size so that all individuals above a certain size are gaudy males (Fig. 9.8). There is no evidence that sex reversal is inhibited by social dominance.

Robertson & Choat (1974) described a social system of this sort in the small Indo-Pacific wrasse *Thalassoma lunare*. This common species occurs as a spread of individuals inhabiting large overlapping home ranges. Fish forage singly or in small groups within their ranges. Drab adults are green with blue bellies and pink heads, gaudies are altogether brighter with a blue cast over most of the body. Sexual activity is restricted to a short time just after high tide each day and during this period gaudy males temporarily defend reproductive territories, displaying aggressively and sexually within them. The sexual display consists of short bursts of rapid flutterings of the bright blue pectoral fins, brief bursts of body vibration and occasional upward dashes. If this attracts a drab gravid female the male intensifies his display and the pair may then perform a spawning rush – upwards for several metres, then an abrupt turn as the eggs and sperm are released in a small cloud, then down more slowly, separating as they descend. Aggregated spawning may also occur in which a number (sometimes several hundred) of drab but sexually mature primary males surround a gravid female and a mass spawning rush ensues at the peak of which the female sheds her eggs and a large cloud of sperm is released by the numerous males.

A similar system has been described in the Caribbean striped parrot fish *Scarus croicensis* except that in this species there is a greater variety of behavioural phases and colour patterns (Warner & Downs, 1977). Drab individuals have longitudinal dark brown stripes but larger drab females have bright yellow pelvic fins; gaudy males are bright blue. The behavioural phases include stationary groups inhabiting undefended home ranges of up to 50

Fig. 9.8. Sexual composition of successive length classes of the striped parrot fish *Scarus croicensis* showing sex reversal from female to secondary male at a body length of about 10 cm. 1°, primary; 2°, secondary. After Warner & Downs (1977).

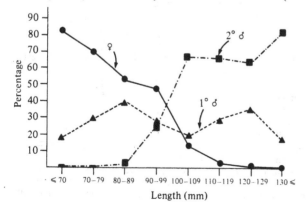

m^2, two types of mobile foraging groups, and small territorial groups defending feeding and reproductive areas of 10–12 m^2. Table 9.1 shows the composition of these different social groups and Fig. 9.9 shows the size range in each category. It may be seen that the highest proportion of sexually inactive (small, immature) fish are found in the stationary groups. From these groups the maturing females move to foraging groups of small females, and into territories. After further growth these females change sex and join all-male foraging groups as gaudy secondary males. Maturing primary males move directly from the stationary groups to the all-male foraging groups where, after further growth, they too become gaudy. Finally the larger gaudy males, both primary and secondary, move into territories.

These territories are unusual since although they contain a single large gaudy male and several females, they are defended mainly by the dominant female (Buckman & Ogden, 1973). This female defends her territory against conspecifics using an aggressive display of bright yellow pectoral fins and by highly aggressive mouth to mouth interactions; but she tolerates a few subordinate females within her territory. The male's defence is restricted to chasing off other conspecific males. As in *T. lunare* pair spawning

Table 9.1. *Percentage sexual composition of behavioural groups of striped parrot fish* Scarus croicensis *(for further details see the table footnote and also the text)*

	Behavioural group[a]					
Behavioural group	Inac. 1°♂ drab	Ac. 1°♂ drab	Inac. ♀ drab	Ac. ♀ drab	Ac. 1°♂ gaudy	Ac. 2°♂ gaudy
Stationary	19	26	29	19	3	4
Foraging 1[b]			50	50		
Foraging 2[b]		11			37	52
Territorial drab			7	93		
Territorial gaudy					20	80

Data from Warner & Downs (1977)
[a] Ac., sexually active; Inac., sexually inactive; 1°, primary; 2°, secondary.
[b] Foraging group 1 (all female) contained fish 5–9.5 cm long whereas in foraging group 2 (all male) the fish were 9–13 cm long, see Fig. 9.9.

occurs within territories and occasional aggregated spawning also takes place.

9.3.5 Reproductive strategies

The protogynous social systems described above represent various balances between the reproductive advantages of each sex. Eggs are larger than sperm and are metabolically more expensive to produce. It is therefore reproductively more profitable to be a male, but only if the population structure allows the fertilization of several females. Suitable population structures occur in many species of animals and the problem that tends to arise is that large males monopolize all the females leaving small males with no reproductive role. In the reef fish discussed above two solutions are adopted. The first is to reproduce as a female while waiting to grow big enough to be successful as a male. The second is to be a male from the outset, but to grow quickly so as to spend as little time as possible as a small male – there is evidence that primary males do grow relatively fast and become gaudy sooner than secondary

Fig. 9.9. Mean lengths of striped parrot fish *Scarus croicensis* in each behavioural group. S, stationary; F, foraging; T, territorial; 1°, primary; 2°, secondary. Points are means with 95% confidence limits and ranges. For further details see Fig. 9.8, Table 9.1 and text. After Warner & Downs (1977).

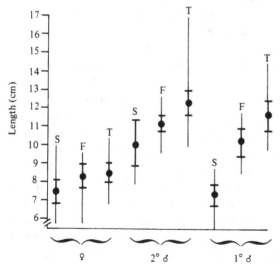

males (Warner & Downs, 1977). In *Labroides* and *Anthias* the first solution is adopted but in *Thalassoma* and *Scarus* there is a mixture. This persists because the two solutions balance each other: being a female entails slower growth because of the investment in eggs, thus, provided there are sufficient females, there is a profit to be made by being a fast-growing male. Too many fast-growing males, however, would lead to excessive competition for the few remaining females and it would then become more profitable, on balance, to be a female instead. A further consideration is the possibility that mature drab primary males may obtain some reproductive profit from aggregated spawning. Observations suggest, however, that mass spawning is much less common than pair spawning and, in any case, the number of eggs fertilized by any one male in the swarm must be rather few. Aggregated spawning is thus probably only marginally profitable.

Protogyny and rapidly growing primary males are not, of course, the only solutions to the problems of reproduction. Many reef fish have what we would regard as normal sexes and a few are protandric. Examples of the latter include species of the clownfish *Amphiprion* (Pomacentridae) which live in symbiotic association with large anemones (see below §9.4.2) (Fricke & Fricke, 1977). Clownfish, so called because of their bright orange, brown and white bars and their impudent behaviour, normally live in pairs, one pair per anemone. Pairs are monogamous and consist of a large female and a smaller male; the female aggressively dominates the male and both fish vigorously defend their anemone (territory) against all intruders. If the female is removed the male changes sex and is joined by a smaller male.

Having accounted for protogyny above, how can a sensible advantage be found to explain protandry? The answer lies in the remark made at the beginning of this section about social systems being adaptations to ecological roles: the appropriate system is the one that accords best with the way of life. In the protogynous species considered so far the distribution of food resources and the methods of feeding result in the populations being distributed so that the ranges of individuals overlap. Thus the chances of a male encountering several females are fairly high and the dominant male strategy pays off. The distribution of clownfish, however, is governed by the distribution of anemones, and these are usually widely scattered with distances of several to many metres between neighbours. This makes it dangerous for fish to move between anemones and since each anemone is usually only big enough to support two fish the reproductive output of the pair is limited by the number of eggs the female can produce. Since big females produce more eggs it pays to be a large female, hence protandry.

9.3.6 Territories

The concept of the territory as a defended area, distinct from the home range which is undefended, has occurred several times in the above discussions. Reef fish almost always have home ranges or territories, and frequently both – in which case the territory is a smaller area towards the centre of the range. Even large open-water fish such as barracudas may sometimes attach themselves to particular parts of a reef and the same individual will be seen in the same area day after day.

Factors which govern territorial behaviour include feeding, reproduction and shelter: territories may be feeding areas, breeding areas, hiding places, or some combination of these. We have seen above (§ 8.6) that territorial damsel fish use their territories as feeding and breeding areas; these damsel fish are unusual in having benthic eggs which are laid within the territory of the male and guarded by him until they hatch. Damsel fish defend their territories against all comers, especially herbivores and egg-eating fish (Ebersole, 1977). According to Itzkowitz (1974, 1977) damsel fish tend to occur in single-species groups of territories. This tendency to colony formation may be related to habitat selection, but it also carries two social advantages. It allows group defence against raiding parties of parrot fish and surgeon fish, and it allows reproduction between near neighbours, removing the necessity for individuals to leave their homes unguarded while they spawn. Group defence and spawning with near neighbours have both been recorded.

On an individual basis the territory size, and vigour with which it is defended, vary from one species of damsel fish to another. Thresher (1977) described several examples of this (Fig. 9.10) and related the differences to the distribution of resources which the territory represented. Any particular strategy is presumably an evolved answer to the problems posed by the following type of questions: is the resource worth defending? How large a part should be defended? Since defence is energy consuming and dangerous what is the minimum vigour required to protect the resource? A clutch of eggs is a small area worthy of vigorous defence, but if the resource is a fast-growing alga one might expect a larger, less vigorously defended territory.

Rather different territories are those defended just against conspecifics. These include the territories of the cleaner fish *Labroides* and those of the territorial phases of some parrot fish and wrasse.

Fig. 9.10. Territory size and territory defence in three species of Caribbean damsel fish. (*a*) *Eupomacentrus partitus* vigorously defends small territories; (*b*) *E. planifrons* vigorously defends large territories; (*c*) *E. variabilis* weakly defends large territories. Experimental procedure was to place a live 'intruder' in a glass jar at a measured distance from the territory centre and to count the number of bites/min directed at the imprisoned intruder through the glass walls of the jar by the territory holder. Each point shows the mean and range of several experiments. From Thresher (1977).

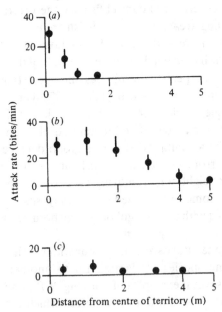

In the cleaner fish the resource is a location on the reef, such as a conspicuous coral head, where other reef fish come to be cleaned (see below, § 9.4.1). Clearly the interests of the cleaner are served by encouraging customers (other species) and chasing away competitors (conspecifics). In contrast, the food resources of most wrasse and parrot fish (small invertebrates and algal turf respectively) are widely distributed and more appropriately harvested within home ranges since the foraging areas are too big to be effectively defended. Territories in these species are thus normally temporary display sites for large males, defended only against other males. In the striped parrot fish *Scarus croicencis*, however, females in the territorial phase defend permanent territories against conspecifics only. Ogden & Buckman (1973) observed that this species migrated diurnally from shallow day-feeding areas to deeper night-resting areas (parrot fish spend the night resting in coral crevices encased in a thin mucous bag – the mucus is secreted by the fish and may make it more difficult for night hunting predators such as moray eels to detect the fish inside), but territorial females rest at night within their territories (Buckman & Ogden, 1973). This suggests that the territory may function partly as a night shelter, the advantage being that the possibly dangerous migration can be avoided. Damsel fish also use their territories as night-resting sites, and those butterfly fish which range widely during the day in monogamous pairs often show territorial behaviour only at dusk during the brief defence of a regular night-resting site (Ehrlich *et al*. 1977).

9.4   Symbiotic relationships

There is a wide variety of symbiotic and commensal relationships to be found on coral reefs (Fricke, 1975). Some have already been described (§8.7, 9.2.1, 9.3.1) and others mentioned; the reef corals themselves are a major example (§ 7.2). In this section I want to enlarge on cleaner fish and clownfish since these associations are particularly striking and have received considerable attention.

9.4.1   Cleaning symbiosis

An early description of cleaning symbiosis was that of Limbaugh (1961) who noticed that certain fish and shrimps would often associate with other larger fish (hosts) and would move over and around them, even venturing under their gill covers and into their mouths, apparently removing and eating ectoparasites and necrotic tissue from their skins. This was observed in both temperate waters and on coral reefs. Later research has shown the behaviour to be highly developed in coral reef communities. Cleaner fish generally have regular stations on the reef to which the hosts come and at which cleaning occurs. Cleaner shrimps tend to lurk in caves and crevices and clean fish which shelter inside. Limbaugh (1961) suggested that cleaner fish are very important members of the community since they keep other fish free of ectoparasites and external infections. He cited an experiment in which the removal of all cleaner fish from a small reef resulted in substantial emigration of fish and clear signs of infection on those that remained. Since that time several attempts have been made to repeat this experiment but in no case have such dramatic results been obtained (Gorlich *et al.*, 1978). It has been found that the guts of cleaning fish do indeed contain ectoparasites and epidermal material but, perhaps surprisingly, there is no clear evidence from removal experiments that cleaners control the levels of ectoparasitic infection in reef fish populations. Nor is there evidence that removal of cleaners causes resident fish to move elsewhere. Perhaps the experiment normally gives negative results because of the difficulty of removing all cleaners from a reef: the specialized adult cleaners are relatively easy to remove but juvenile cleaner fish and cleaner shrimps are often cryptic in their habits. In addition there are a number of part-time cleaners, e.g. juvenile angel fish (Pomacanthidae), which may increase their cleaning behaviour in the absence of the specialists. However, although particular cleaning species may not be as indispensable to the community as Limbaugh (1961) suggested, it cannot be doubted that both cleaner and host derive benefit at an individual level from cleaning behaviour.

Turning to the behaviour itself, this has been studied in several different parts of the world. One study was that of Potts (1973) in Aldabra on *Labroides dimidiatus* (Fig. 9.7). Potts watched numerous cleaners, both adults and juveniles, and recorded the frequency of cleaning bouts and the details of cleaning behaviour.

His data on bout frequency (200 hosts cleaned per hour by an adult *Labroides*) give a good idea of the industrious nature of these fish, and his data on the behaviour were condensed into an ethogram reproduced in Fig. 9.11. Various points in this ethogram deserve amplification. In the approach phase the cleaner may perform an undulating display which, in combination with the conspicuous colour pattern of the cleaner, may serve to attract potential hosts and elicit invitation postures. These host postures are often unusual and characteristic. The creole wrasse *Clepticus parrae* in the Caribbean adopts a head-down posture with the body held between 45° and vertical; it is not uncommon to see several of these

Fig. 9.11. A diagram illustrating the interactions between the cleaner wrasse *Labroides dimidiatus* and a host fish. For further details see text. From Potts (1973).

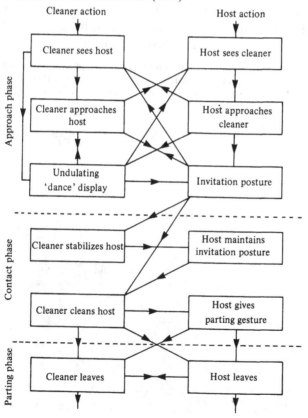

fish posing together at a cleaning station, the cleaner working over one of them while the others await their turn, as if in a queue (Fig. 9.12). Potts (1973) observed that contact with the cleaner appeared to stabilize the host and make it more likely to continue to pose (i.e. less likely to swim away); sometimes a cleaner was seen to touch and stabilize several hosts before proceeding to clean them one by one. Potts also noticed that some species of hosts were preferred to others and that particular species were cleaned in characteristic ways. In goatfish, for instance, the gills were more frequently cleaned than any other part of the body whereas in snappers the tail and flanks were given most attention and the head was apparently avoided. Photographs of cleaner fish in action often show the mouths and gills of predators such as snappers and groupers being cleaned; this is because such scenes are at once spectacular and incongruous. Potts' observations suggest that they may also be relatively rare probably because there is a real danger involved in cleaning near a predator's mouth. At the end of a

Fig. 9.12. Creole wrasse, *Clepticus parrae*, posing at a cleaning station. The cleaner (arrow) is attending to the fish on the left. The creole wrasse are each about 15 cm long. The small fish on the right is a damsel fish. 10 m, Jamaica.

cleaning bout either host or cleaner may simply depart, or the host may shake itself, possibly signalling its intention to leave.

A final interesting point about cleaners is that a colour pattern similar to that of *Labroides* (Fig. 9.7) is found in other unrelated cleaners, for instance in the Caribbean neon goby *Elacatinus*. The longitudinal dark stripe may almost be viewed as a professional uniform. It has even been adopted by an aggressive mimic, the sabre-toothed blenny *Aspidontus*, which approaches larger fish and bites pieces from their fins.

9.4.2 Clownfish

Clownfish bathing amongst the tentacles of their anemones are one of the most striking examples of interspecific association to be seen on coral reefs. Clownfish are found in the Indo-Pacific region and comprise several species. Adults are between 5 and 12 cm long and their anemones are 15–50 cm or more in diameter. Some species of clownfish are species specific in their association with anemones while others are able to live with several different species (see Mariscal, 1972, for review). The bright colours of the clownfish, their aggressive territorial behaviour, and the final retreat in which the fish is lost to sight among the tentacles of its anemone, all combine to draw the attention of the diver. How do the fish avoid being stung? What benefits do the partners derive from the association?

The answer to the first question is that the fish arrange protection for themselves: there is no evidence of altruism on the part of the anemone. The protection consists of a thick external coat of mucus which is acclimated to contact with the anemone. Acclimation is lost if a clownfish is separated from its anemone for a day or two. Upon reintroduction a deacclimated fish does not dive directly into its anemone as it might normally do, it approaches more slowly and dips its lower fins and tail in among the tentacles. The latter adhere to the fins by the firing of nematocysts and the tentacles contract, bending towards the mouth as in normal feeding behaviour; but the clownfish easily breaks free and comes back for more. This repetitive dipping in and breaking free constitutes acclimation behaviour and, as it proceeds, the clownfish dips deeper and deeper amongst the tentacles. Eventually the tentacles cease reacting and acclimation is complete. Time to completion varies with the species of anemone from 10 minutes to 45 hours. It is not known precisely what happens to the mucous coat during acclimation; probably something is obtained from the anemone, such as discharged nematocysts or additional mucus, which makes it impossible for the anemone to distinguish between the fish and its own tentacles. Acclimation is not in the anemone since acclimated fish retain acclimation if transferred to another anemone of the same species but is in the mucous coat since fish immediately lose acclimation if the coat is removed.

The main service performed for the clownfish by its anemone is protection. Nestling among the tentacles the fish is quite safe from predators. Additional possible benefits to the fish include the ingestion of waste materials and mucus from the oral disk of the anemone and occasional ingestion of tentacles. However, the main food of most clownfish is zooplankton taken from the water above the anemone. The benefits to the anemone from the association are less obvious but again involve feeding and protection. Mariscal (1972) described clownfish feeding their anemones with large objects which came drifting past, and occasionally seizing small fish and thrusting them in amongst the tentacles. Protection of the anemone results from the aggressive territorial behaviour of the clownfish which deters potential anemone predators. This type of benefit to the host is probably fairly common on coral reefs; it has been mentioned above in commensal coral crabs which deter the coral predator *Acanthaster* (§ 8.7) and has been recorded for a snapping shrimp which defends its anemone host against the predatory worm *Hermodice*.

9.5 Resource partitioning and diversity

The view of coral reefs as orderly systems predicts strict resource partitioning by species. Important resources include food and shelter and these two are combined in the resource of space. It should be clear from what has already been said in this chapter that there is much overlap between species of reef fish in the use of resources. The different feeding categories each contain many species feeding – sometimes in mixed schools – on similar food; individual shelter sites, such as caves, may be used simultaneously by several species; the space on the reef is an overlapping maze of home ranges and territories inhabited by a wide variety of species. The usual way of explaining this apparently chaotic situation is to suggest that there are always small but critical differences in feeding behaviour or space utilization which define separate ecological niches and allow the species involved to coexist in conformity with Gause's hypothesis (the competitive exclusion principle: the number of species is no greater than the number of ecological niches). Thus clownfish and *Chromis* both feed on zooplankton but one shelters in anemones while the other shelters in branching coral. A particularly nice example of resource partitioning is revealed by diving at night (Luckhurst & Luckhurst, 1978) when it is observed that the 'day-shift' of wrasse, damsel fish, parrot fish, etc. are replaced by a 'night-shift' of squirrel fish (Holocentridae) and cardinal fish (Apogonidae). Night-shift fish usually have large eyes and shelter in caves or crevices during the day, squirrel fish feed on invertebrates, especially polychaete worms, and cardinal fish are planktivorous. Another good example of resource partitioning is that, within a species, adults and juveniles often have different habitats.

In spite of a general tendency to resource partitioning, however, apparent overlap between ecological niches has been observed – not so much within a small area as between neighbouring areas. Thus having assigned a particular species to an ecological niche on the basis of a spatially limited study, one may find in an adjacent area a different species occupying an apparently similar niche. The result is that within the same general reef area there may be more species present than there are identifiable ecological niches. Sale (1980) suggested that this apparent conflict with Gause's hypothesis follows from the existence of random factors associated with recruitment. He suggested three filters to recruitment: first, the general suitability of the physical habitat, second, the availability in the plankton of potential recruits (almost all reef fish have a planktonic larval stage), and third, the interactions of recruits (e.g. competition, predation) with prior residents. The second filter is subject to considerable variation – larval availability in the plankton is notoriously unpredictable. The third filter is also unpredictable since the identity and numbers of prior residents depend on original chance recruitment and on subsequent chance disturbance. As has been stressed above, coral reefs are by no means immune from disturbance. Thus reef fish communities, like sessile communities on rocks (§ 3.3) and communities of reef corals (§ 8.8), are dynamic assemblages; responsive to chance and so more or less unstable, expecially at the local level. The communities contain both specialist and opportunist species, the latter possessing a degree of niche flexibility. This model is in line with Huston's (1979) general hypothesis on community diversity and describes a state in which order and chaos are balanced by disturbance and chance.

Part IV: Level Substrates

10

Level substrates

10.1 Introduction

Level substrates are usually composed of mud, sand, gravel or a mixture of these, the characteristic particle size depending to a great extent on the prevailing water movement. Currents transport particles, either in suspension or by rolling them along the sea bed; when the currents slow down these particles may be deposited. Clearly it requires fast currents to move heavy particles like gravel, whereas fine silt can be carried by quite slow currents. Thus one often finds a correlation between the size of the particles making up the substrate and the strength of the prevailing currents: a muddy bottom usually indicates weak currents while gravel or cobbles indicate strong currents. This correlation is only upset by particle availability. If, for instance, an area receives a superabundance of sand from some distant source where erosion is occurring, then the bottom is likely to be sandy whatever the strength of the current – as fast as the sand is carried away more sand arrives. Similarly, an area with a lot of mud in suspension will develop pockets of mud on the sea bed despite fairly strong currents because there are always sheltered spots where mud can settle, such as behind a rock, or times of slack water when mud can be deposited. Currents, by transporting particles at one speed and depositing them at another, have the ability to accumulate certain substrates in particular places. Banks of gravel and sand and the muddy bottoms of some bays are examples.

On a smaller scale the shape of the bottom may also be due to the action of currents. Ripple marks on sand or gravel are shaped by currents in the same way as the wind shapes sand dunes. These ripple marks are usually asymmetrical in section with a gentle windward slope and a steeper leeward slope, indicating that they are moving slowly down current. Ripple marks formed by the to-

and-fro movement of waves, however, are symmetrical. Storm waves can form quite large ripples in coarse sand and gravel which may persist for months after the storm has passed (Fig. 10.1). Some storm-generated ripples found at 50 m in the English Channel measured 125 cm from crest to crest and each ripple was 20–25 cm high (Flemming & Stride, 1967). Such features serve as reminders to the diver of environmental conditions that are unlikely to be observed directly.

The fauna of level substrates, the 'benthos' of most marine biology texts, has been extensively studied by techniques other than diving. Unlike the case in rocky environments it is relatively easy to sample the substrate by scooping some up in a dredge or grab lowered from a boat. The wide use of these techniques has led to detailed descriptions of community composition to which I will only refer briefly here (see Mills, 1969, for review). The macro-fauna is conveniently divided into the infauna, which burrows, and the epifauna which lives on the surface. Characteristic communi-

Fig. 10.1. Wave-induced ripples on a coarse sand/fine gravel substrate at 15 m off the Chesil Bank, English Channel. The irregular pits in the foreground are the result of bioturbation. The fish are bib (*Gadus luscus*). Distance between ripple crests is 50–75 cm.

ties are found to inhabit the various grades of substrate and the bulk of the animals are suspension or deposit feeders. Suspension feeders, e.g. many of the burrowing bivalve molluscs, are dominant in the coarser grades of deposit since these environments tend to be exposed to faster water movement. In fine sediments deposit feeders are usually dominant and include other burrowing bivalves, worms, heart urchins and holothurians. The epifauna, if sedentary, usually suspension feeds and passive suspension feeders such as some brittle stars may be common on current-washed bottoms. Active suspension feeders such as clumps of mussels, slipper limpets or ascidians may be found on muddy bottoms where a stone or shell has provided an initial settlement point. Mobile epifauna such as crabs and fish are usually predatory. A good example of direct description of a series of benthic communities correlated with decreasing current strength and substrate particle size is to be found in Erwin's (1977) account of a diving survey of a sea-loch in Northern Ireland.

Dredges and grabs continue to be used to study the benthos of deep, offshore areas and very turbid environments, but in suitable places diving is increasingly becoming an important method. The troubles with grabs and dredges include sampling variability, lack of precision in placement of samples, and the inability to observe directly. Sample variability results from the lack of control over the depth of bite or dig of grab or dredge, especially in firm substrates like gravel. A diver, however, can see or feel the depth in the substrate at which the sampler is operating and can often see how much of the epifauna escapes. A device used by divers to sample the benthos is the air-lift suction sampler which consists of a vertical pipe into the bottom of which compressed air is introduced. The air rising up the pipe creates sufficient suction to excavate sediment and to lift the benthos up the pipe to be caught in a bag at the top (e.g. Christie & Allen, 1972). More unique contributions of diving, however, have come from the ability to place samples precisely and to observe directly. The subjects discussed in this chapter derive from these abilities and include a brief description of sea grass communities, studies on dispersion patterns, studies on 'islands' of hard substrate, and work on burrowers and their burrows.

10.2　Sea grass beds

Sublittoral level substrates do not normally support conspicuous primary producers (where there is sufficient light benthic diatoms may be numerous), but beds of sea grass are exceptions. Sea grasses are able to live on particulate substrates because, being flowering plants, they have roots which anchor them to the sand or mud. The tangle of roots and rhizomes binds the substrate and the forest of narrow green strap-like leaves up to 30 cm long promotes sedimentation so that a sea grass bed often forms something of a bank raised slightly above the level of the surrounding sand. Sea grass may be exposed at low spring tides and in very clear water can extend down to 30–40 m; they are most common, however, in 1–5 m. Heavy wave action uproots the plants and grass beds do not develop in places subject to strong water movement, they are commonest in shallow bays and lagoons. Common sea grass genera include *Zostera* in temperate regions and *Posidonia* and *Thalassia* in subtropical and tropical environments.

Grass beds are important because, apart from supporting rich communities, they produce a large amount of detritus – 40–60% of production – in the form of broken leaves which drift away in the current. This, like the detritus produced by kelp forests, contributes to the nutrition of suspension and deposit feeders in other communities. The remainder of the production is grazed by herbivores, the most important of which are often sea urchins, but fish and crustaceans may also bite pieces from the leaves (Fig. 10.2). Some species of urchin prefer the green growing parts but others feed mostly on the distal ends of the ribbon-like leaves where the plant tissue is largely dead. Such leaf tips bear a rich encrusting community – algae, hydroids, bacteria, ciliates, etc. – and are probably just as nutritious as the bright green, less heavily colonized, proximal parts of the leaves. Measures of net primary production in sea grass beds are often in the region of 1 kg C m^{-2} yr^{-1}, of which a substantial fraction may be produced by epiphytic algae (Mann, 1982). This level of productivity is similar to that found in kelp forests.

The substrate in which sea grass is rooted harbours an infauna of bivalve molluscs and annelids and, on the surface, a partly resident and partly sheltering epifauna of sea urchins, fish, hermit crabs and deposit-feeding holothurians. Large gastropods may also be present, especially in the tropics. Strombid gastropods or conchs are lumbering tropical herbivores which lever themselves along using the spiked operculum on the end of the muscular foot. The shell of the queen conch *Strombus gigas* from the Caribbean may be 20–30 cm long and the animal feeds on epiphytic algae scraped from the leaves of sea grass. Large strombids are commercially important since their flesh is often a local delicacy (conch soup, conch fritters) and their large and beautifully coloured shells can be sold to tourists. In some areas overcollection has drastically reduced the populations of these previously common and useful animals. Another large gastropod which occurs in Caribbean sea grass beds is the predatory helmet shell *Cassis* spp. These snails vary from 10–30 cm long and feed on sea urchins. They squirt a toxic saliva over the urchin to immobilize the spines and then rasp a hole in the test with the radula which lies at the end of a long proboscis. Other in-

Fig 10.2. Diagram illustrating the production and consumption of blades of the sea grass *Thalassia testudinum*, mainly by the urchin *Lytechinus variegatus*, in Kingston Harbour, Jamaica. Units are g dry weight /m²/week unless otherwise stated. Modified from Greenway (1976).

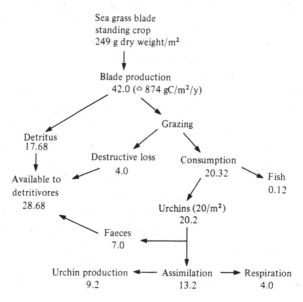

teresting animals that feed extensively on sea grass include green turtles and manatees. The diversity of the epifauna of a sea grass bed is dramatically increased by the chance occurrence of hard substrate. This can take the form of a clump of mussels or a coral boulder and the extra diversity results from the provision of a whole new range of habitats.

10.3   Dispersion patterns
Individuals in a population can be distributed at random in their environment, or they can be evenly spaced, or aggregated into groups. Level substrates, unlike rocky substrates, are often relatively uniform over wide areas and are therefore good places to study dispersion patterns. Small scale environmental variability does not intervene to obscure the patterns nor to explain them trivially by habitat selection. Patterns found on level substrates are more likely to be genuine biological entities. Studies of dispersion clearly require precision in the placement of samples since comparisons between adjacent samples must be made. Such studies have shown that aggregation is a common pattern and that even distributions are rather rare.

Infaunal animals sometimes occur at very high population densities but it is difficult to make out whether their dispersion departs from random without digging up the bottom. This was done in a Norwegian fiord by Angel & Angel (1967) who took a series of systematically positioned core samples on a muddy substrate at 35 m (Fig. 10.3). They found a variety of infaunal animals including annelids, foraminiferans, holothurians, bivalve molluscs and hydroids and most of the species that occurred in sufficient numbers for dispersion analysis showed aggregation. Some aggregation patterns were quite complex: the polychaete *Owenia*, for instance, occurred at random within aggregations 25–50 cm across and these aggregations themselves were randomly distributed within larger aggregations several metres across. It is very difficult to account for dispersion patterns such as these; microhabitat differences are often suggested as possible explanations but grain size analysis, the commonest procedure for investigating the physical habitat, is a fairly crude method for seeking out ecological niches – especially when the tolerances of the species in question are poorly known. The trouble is that one cannot see what is going on under the surface of the sediment.

Much more is known of epifaunal dispersion patterns because one can see the animals and often see what they are doing. The best known are probably the aggregations of brittle stars of the genus *Ophiothrix* (Fig. 10.4) (Warner, 1971; Warner & Woodley, 1975). Several species of this genus form dense aggregations below about 15 m deep on current-washed level bottoms of mixed gravel with rocky outcrops. Population density within these aggregations can reach 3000 m^{-2}. The sea bed appears as a tangled mass of brittle stars but the edges of aggregations are quite sharp and there always appears to be plenty of space into which individuals could disperse if they wished. That they do not so wish was shown by experiments carried out underwater on *Ophiothrix fragilis* off the south coast of England (Broom, 1975). Individuals were removed from an aggregation and were placed on a bare area nearby. It was found that

Fig. 10.3. Sampling procedure adopted to detect dispersion patterns in infaunal organisms; a surface supply of air was found to be convenient in this case. After Angel & Angel (1967).

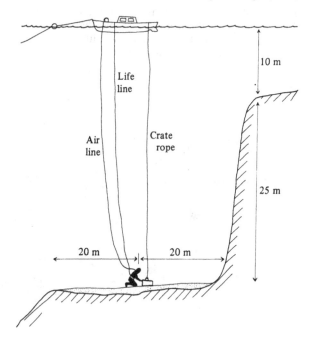

they would walk across current until they encountered other brittle stars (Fig. 10.5). They would then climb in amongst the others and stop walking. No other encounter (with a rock or a hydroid clump for instance) would stop them; the response to stop was specific to members of their own species and is therefore a social response. Thus these aggregations are maintained in the short term by adult social behaviour. In the long term there is evidence that larval behaviour is important: larvae appear to settle from the plankton onto the bodies of adults and areas of brittle star aggregation are on record as persisting for at least 50 years. These brittle stars are suspension feeders; they stretch their arms up into the current and catch particles on their sticky tube-feet; when the current speed increases, however, they link arms and crouch down close to the substrate (Warner & Woodley, 1975). In this state the aggregations form stable mats which are not dislodged and carried away by the current (as isolated individuals are). Stability in currents is probably the main advantage of aggregation in *O. fragilis* but there are other possible advantages. First, there is the advantage of suc-

Fig. 10.4. A dense aggregation of the brittle star *Ophiothrix fragilis*. The gap in the aggregation shows that there is space available for dispersal. The arms of the brittle stars are about 7 cm long. 15 m, Torbay, English Channel.

cessful fertilizatation during the breeding season. Second, neighbouring brittle stars prop each other up, thus fewer arms are needed to support the body and more can be extended into the current for feeding. Third, it is possible that the forest of extended arms slows down the current and promotes deposition of particles (food) within the aggregation. One might think that the crowded nature of the aggregations would increase competition for food but it is likely that because of the last two advantages individuals gain more food, not less, by aggregating (Warner, 1979). A final possible advantage of living in an aggregation is that it may give protection from predation. A predator attacking a brittle star aggregation is more likely to get hold of a mouthful of spiny arms than of juicy bodies since to single out and attack a particular individual may be difficult because of the interference from the writhing tangle of neighbouring arms.

Fig. 10.5. Example of the route of a walking brittle star (*Ophiothrix fragilis*) after being removed from an aggregation and placed in isolation on muddy gravel nearby. Initial walk, longest walk and overall direction were usually at about 90° to the current (arrow). The pie diagram shows the results for overall direction from 25 experiments, the numbers refer to the number of individuals walking in that direction. Walking direction was random with respect to direction of origin or nearest aggregation. After Broom (1975).

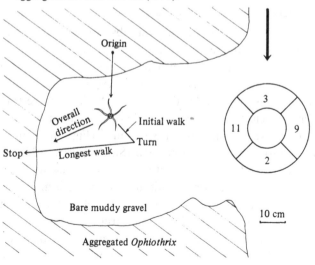

A rather different habitat, studied by Fager (1968) in California, was a sand substrate between 5 and 10 m deep. This environment was constantly disturbed by waves which created miniature sandstorms on the bottom as they passed overhead. The area was also subject to rapid temperature changes due to internal waves on the thermocline at this depth. Fager found nine dominant epifaunal species: two anemones, one sea pansy (Pennatulidae), one brittle star, one starfish, three carnivorous gastropods and one hermit crab. Seven of these species showed aggregation and in one, the sea pansy, it was suggested that the groupings served to stabilize the sand in which the sea pansies were anchored; this would lead to greater stability for groups than for individual sea pansies and is a similar advantage to that suggested above for brittle star aggregations. Fager studied this community for several years and was surprised to find that the populations of the dominant animals remained fairly stable from year to year in this apparently unstable environment. A study conducted ten years later in the same area, however, showed that marked changes had taken place in seven of Fager's nine species, and further changes during the course of this later study were also noted (Davis & Blaricom, 1978). These changes were thought to be due to variable recruitment from the plankton.

10.4   Islands of hard substrate
    The very uniformity of level substrates, referred to above, is responsible for the relatively low diversity of these communities: there are relatively few ecological niches. However, where isolated hard substrates occur diversity can increase dramatically. An example of this was found in the northern Adriatic where epifaunal multispecies clumps occurred at a density of about 6 m^{-2} and a mean fresh weight biomass of 370 g m^{-2} (Fig. 10.6) on a level muddy bottom at about 23 m deep (Stachowitsch, 1977). These clumps were sedentary associations of more than one (usually several) sessile species, on and within which various mobile species sheltered. About 90% of the biomass of the clumps was accounted for by suspension feeders. The sessile structural components were mainly sponges and ascidians with epifaunal bivalves and anemones also being common; the most numerous of the mobile

sheltering animals were suspension feeding brittle stars, with holothurians, annelids, crabs, etc. also occurring. Clumps were easily lifted from the substrate and were found to be growing on a base composed of one or more shells or shell fragments. Frequently these shells were gastropod shells showing evidence of previous occupancy by hermit crabs, and living hermit crabs were often found to support smaller but mobile multispecies clumps upon their shells. Stachowitsch (1977) surmised that the large sedentary multispecies clumps originated as smaller mobile clumps on occupied hermit crab shells. Hermit crabs were common in the area and an experiment in which marked, empty shells were placed on the sea bed showed that hermit crabs readily occupied them; marked shells with blocked apertures, however, sank into the mud where they were unavailable for epifaunal colonization. Thus hermit crabs helped to keep shells on the surface where they could be colonized by sessile epifauna. Once established, multispecies clumps appeared to be able to persist and grow by spreading laterally and attaching to small stones or shells encountered on the surface of the substrate.

Another example, in a sea loch in Northern Ireland, was

Fig. 10.6. Appearance of a multispecies clump on a muddy bottom in the northern Adriatic. The clump is formed by a sponge, with an anemone, hydroids, holothurian and brittle-stars in association. Inset shows a smaller mobile clump on a hermit crab shell. Details of these associations are given by Fedra (1977) and Stachowitsch (1977).

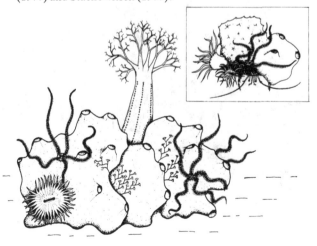

observed by Erwin (1977) and consisted of clumps of large mussels, *Modiolus*, resting on a muddy substrate. The clumps were colonized by other sessile elements including sponges, ascidians, serpulid worms and hydroids, and supported a community of mobile shelterers, predators and scavengers. In this case the

Fig. 10.7. Reef-formation by the serpulid worm *Serpula vermicularis*. (*a*) a piece of serpulid reef; photograph by the University of Reading Photographic Department. (*b*) detail of the serpulid tubes; note encrusting bryozoans; photograph by D. W. J. Bosence.

(a)

10 cm

dominant structural animal, the *Modiolus*, generates additional hard substrate as it grows and dies, in the form of its own shells, so one expects the community to persist even in a depositional environment – the intervention of mobile de-silting agents like the Adriatic hermit crabs is not necessary.

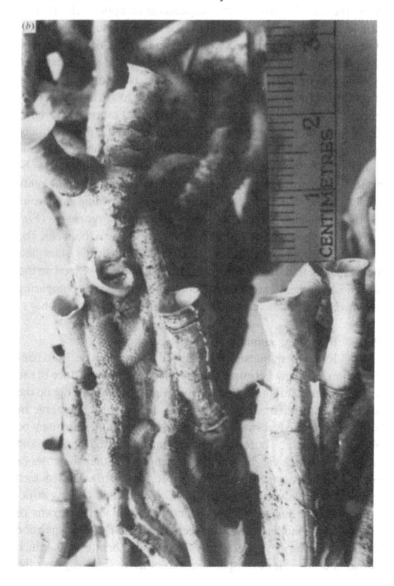

A particularly good example of a community generating its own hard substrate was found between 3 and 19 m in a muddy sea loch or lagoon on the west of Ireland (Bosence, 1979). This was a community of serpulid worms which formed reef-like structures up to 2 m high by settling and growing one on top of another. Fig. 10.7 shows the way the tubes of the serpulids grow upon each other to form large erect structures and Fig. 10.8 shows the suggested mode of development from initial settlement and growth of primary reef on isolated outcrops of hard substrate to growth of secondary reef on collapsed primary reef. Collapse results, as in coral reefs, from the strain on the base imposed by additional growth and from progressive weakening of the base by boring organisms. Bosence speculated that the larvae of this serpulid may settle preferentially on tubes of the adult, but concluded that although this might be so, adult tubes were in any case the most readily available hard substrate. As in the examples described above, the serpulid reefs attract a wide variety of other organisms by providing numerous ecological niches. Coralline algae, hydroids, bryozoans (Fig. 10.7*b*) and sponges encrust the tubes, in crevices between the tubes soft-bodied polychaetes, small crabs and brittle stars find hiding places, and predators such as starfish and wrasse feed on the serpulids. Bosence extended the coral reef analogy by comparing these predators to the crown-of-thorns starfish and parrot fish.

10.5   Burrowing animals

Infaunal animals were dismissed above from detailed consideration by divers because, being buried below the surface of the substrate, they cannot be directly observed and one must dig up the sea bed to find them. In fact, however, many of them emerge or project from the substrate from time to time and can then be observed directly. Those that emerge include burrowing fish and crustaceans which can be watched foraging near their burrows or engaged in burrow maintenance (Figs. 10.9 & 10.10). Those which project are usually sedentary animals which extend feeding structures onto the surface. These include the tentacular crowns of burrowing anemones, the arborescent oral tentacles of suspension feeding holothurians, the arms of burrowing brittle stars which extend into the water to suspension feed or move sinuously over the

Fig. 10.8. Diagrams illustrating the development of serpulid reefs. (*a,b*) reef growth on pre-existing hard substrate; (*c,d*) reef collapse and further growth on collapsed sections. From Bosence (1979).

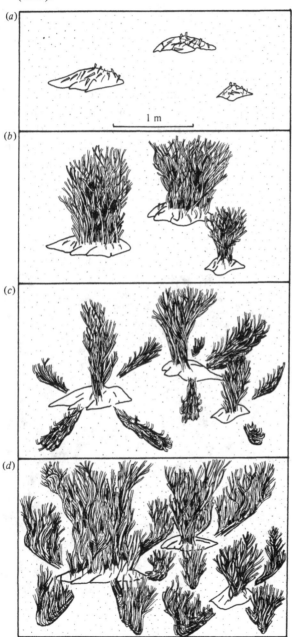

surface of the substrate deposit feeding, the front ends of deposit-feeding holothurians extending over the surface, and the tentacular crowns of suspension- and deposit-feeding polychaetes. As well as being able to watch the behaviour of these animals, divers can use the occurrence of feeding structures and the tell-tale marks in the substrate caused by feeding activities to study population density and dispersion patterns.

The most noticeable signs of infaunal presence are the entrances of burrows. Not all infaunal animals construct burrows in the sense of tunnels through the substrate along which the animals move, many just bury themselves. Bivalve molluscs and heart urchins, for instance, are usually simply embedded in the substrate and only have narrow temporary channels of communication with the substrate surface for their syphons or tube-feet. Infaunal fish and

Fig. 10.9. The Norway lobster or scampi *Nephrops norvegicus* near the entrance to its burrow on a soft mud substrate at 30 m in Loch Corridon, Scotland. Photograph by J. Main.

Fig. 10.10. The planktivorous red band fish *Cepola rubescens* half-way (about 25 cm) out of its burrow on a muddy bottom at about 13 m off Lundy, Bristol Channel. Photograph by R. J. A. Atkinson.

crustaceans, however, often construct elaborate burrows, the
entrances of which are clearly visible to the diver. An excellent
method for investigating these burrows is to fill them with a slow
setting resin poured into the burrow entrance with the aid of a
monster syringe. The resulting cast of the burrow is dug up later by
divers. These casts pick out in detail the elaborate shapes of the
burrows (Fig. 10.11) and sometimes help to identify the burrower
by entombing it in resin so that it is collected with the cast. Some
burrowers, especially some of the burrowing prawns, very rarely
emerge onto the surface and their galleries descend too deep for
them to be collected by conventional methods: the only way to
collect them and thus to link the burrow with an identified owner is
to trap them in resin. An interesting point that has emerged from
burrow cast studies is that, like islands of hard substrate, large
burrows provide a focus for the activities of a variety of species
other than the burrow owner. For instance, subsidiary burrows
constructed by polychaetes or crustaceans may be excavated from
the walls of the main burrow owned by a fish (Fig. 10.11). Perhaps

Fig. 10.11. Outline of a polyester resin cast of a burrow of the red
band fish *Cepola rubescens* taken off Lundy at about 12 m. The
side banches at (*a*) were made by the burrowing crab *Goneplax
rhomboides* while the complex of tunnels at (*b*) were probably
made by a thalassinoidean prawn. Traced from a photograph in
Atkinson *et al.* (1977).

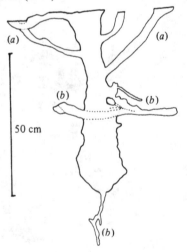

the presence of the large owner gives some protection from predation to the smaller lodgers.

The ability to recognize infaunal animals from the marks they leave on the substrate or from the projecting feeding structures allows divers to survey shallow benthic communities very rapidly. A diver swimming about on the bottom for 10 minutes can examine a very large number of square metres of substrate. Most infaunal communities contain some members which project from the substrate and these species can be used as indicators of the presence of the particular community. Occasional checks on the continued association of the indicator with its community can be made by dredge or grab. Diver-surveys of this sort are best employed in relatively small areas – such as within a bay – where a detailed description of the distribution of communities is required. Such descriptions can be linked to proximity of a source of pollution, or to substrate distribution and hydrodynamic conditions within the bay, (e.g. Warner, 1971). A common pattern is to find coarse to medium mobile sediments close to the shore associated with water movement caused by wave action; these sediments generally contain an impoverished fauna lacking species which build permanent burrows. Further out in deeper calmer water one often finds medium to fine sediments containing such animals as deposit-feeding holothurians and perhaps some burrowing crustaceans. Still further out at the mouth of the bay or near the headlands one may find coarser sediments associated with mainstream tidal currents; here such things as the arms and oral tentacles of suspension-feeding brittle stars and holothurians respectively may be seen projecting from the substrate. Thus if you know what you are looking for, a day's diving may provide more information than a month's tedious boat sampling and laboratory sorting – a moral, perhaps, with which to end: there is no substitute for going and having a look.

REFERENCES

Angel, H. H. & Angel, M. V. (1967). Distribution pattern analysis in a marine benthic community. *Helgoländer wiss. Meeresunters.* **15**, 445–54.

Atkinson, R. J. A., Pullin, R. S. V. & Dipper, F. A. (1977). Studies on the red band fish, *Cepola rubescens. J. Zool. Lond.* **182**, 369–384.

Bak, R. P. M. (1974). Available light and other factors influencing the growth of stony corals through the year in Curaçao. In *Proceedings of the Second International Coral Reef Symposium*, vol. 2, pp. 229–33. Great Barrier Reef Committee, Brisbane.

Bak, R. P. M. (1976). The growth of coral colonies and the importance of crustose coralline algae and burrowing sponges in relation with carbonate accumulation. *Neth. J. Sea Res.* **10**, 285–337.

Bak, R. P. M., Brouns, J. J. W. M. & Heys, F. M. L. (1977). Regeneration and aspects of spatial competition in the scleractinian corals *Agaricia agaricites* and *Montastrea annularis*. In *Proceedings, Third International Coral Reef Symposium*, vol. 1, pp. 143–8. Rosenstiel School of Marine and Atmospheric Studies, Miami.

Barnes, D. J. (1973). Growth in colonial scleractinians. *Bull. mar. Sci.* **23**, 280–98.

Barrales, H. L. & Lobban, C. S. (1975). The comparative ecology of *Macrocystis pyrifera*, with emphasis on the forests of Chubut, Argentina. *J. Ecol.* **63**, 657–78.

Bernstein, B. B. & Jung, N. (1979). Selective pressures and coevolution in a kelp canopy community in southern California. *Ecol. Monogr.* **49**, 335–55.

Bernstein, B. B., Williams, B. E. & Mann, K. H. (1981). The role of behavioural responses to predators in modifying urchins' (*Strongylocentrotus droebachiensis*) destructive grazing and seasonal foraging patterns. *Mar. Biol.* **63**, 39–49.

Birkeland, C., Cheng, L. & Lewin, R. A. (1981). Motility of didemnid ascidian colonies. *Bull. mar. Sci.* **31**, 170–3.

Bosence, D. W. J. (1979). The factors leading to aggregation and reef formation in *Serpula vermicularis* L. in *Biology and Systematics of Colonial Organisms*, ed. G. Larwood & B. R. Rosen, pp. 299–318, Systematics Association Special Volume No. 11. Academic Press, London & New York.

Breen, P. A. & Mann, K. H. (1976). Destructive grazing of kelp by sea urchins in eastern Canada. *J. Fish. Res. Bd. Can.* **33**, 1278–83.

Broom, D. M. (1975). Aggregation behaviour of the brittle-star *Ophiothrix fragilis*. *J. mar. biol. Ass. UK*, **55**, 191–7.

Buckman, N. S. & Ogden, J. C. (1973). Territorial behaviour of the striped parrot fish *Scarus croicensis* Bloch (Scaridae). *Ecology*, **54**, 1377–82.

Buddemeier, R. W. & Kinzie, R. A. III. (1976). Coral growth. *Oceanogr. mar. Biol.* **14**, 183–225.

Buss, L. W. & Jackson, J. B. C. (1979). Competitive networks: nontransitive competitive relationships in cryptic coral reef communities. *Am. Nat.* **113**, 223–34.

Cameron, A. M. (1977). *Acanthaster* and coral reefs: population outbreaks of a rare and specialized carnivore in a complex high-diversity system. In *Proceedings, Third International Coral Reef Symposium*, vol. 1, pp. 193–9. Rosenstiel School of Marine and Atmospheric Studies, Miami.

Christie, N. D. & Allen, J. C. (1972). A self contained diver-operated quantitative sampler for investigating the macrofauna of soft substrates. *Trans. R. Soc. S. Afr.* **40**, 299–307.

Darwin, C. (1845). *The Voyage of the Beagle*. Dent, London.

Davies, P. J. (1977). Modern reef growth – Great Barrier Reef. In *Proceedings, Third International Coral Reef Symposium*, vol. 2, pp. 325–30. Rosenstiel School of Marine and Atmospheric Studies, Miami.

Davies, P. Spencer (1977). Carbon budgets and vertical zonation of Atlantic reef corals. In *Proceedings, Third International Coral Reef Symposium*, vol. 1, pp. 391–6. Rosenstiel School of Marine and Atmospheric Studies, Miami.

Davis, N. & Blaricom, van G. R. (1978). Spatial and temporal heterogeneity in a sand bottom epifaunal community of invertebrates in shallow water. *Limnol. Oceanogr.* **23**, 417–27.

Dayton, P. K., Robilliard, G. A. & Paine, R. T. (1970). Benthic faunal zonation as a result of anchor ice at McMurdo Sound, Antarctica. In *Antarctic Ecology*, vol. 1, ed. M. W. Holdgate, pp. 244–58. Academic Press, London.

Dustan, P. (1975). Growth and form in the reef-building coral *Montastrea annularis*. *Mar. Biol.* **33**, 101–7.

Ebersole, J. P. (1977). The adaptive significance of interspecific territoriality in the reef fish *Eupomacentrus leucostictus*. *Ecology*, **58**, 914–20.

Ehrlich, P. R., Talbot, F. H., Russell, B. C. & Anderson, G. R. V. (1977). The behaviour of chaetodontid fishes with special reference to Lorenz's poster coloration hypothesis. *J. Zool. Lond.* **183**, 213–28.

Endean, R. (1977). *Acanthaster planci* infestations of reefs of the Great Barrier Reef. In *Proceedings, Third International Coral Reef Symposium*, vol. 1, pp. 185–91. Rosenstiel School of Marine and Atmospheric Studies, Miami.

Erwin, D. G. (1977). A diving survey of Strangford Loch: the benthic communities and their relation to the substrate – a preliminary account. In *Biology of Benthic Organisms*, ed. B. F. Keegan, P. O'Céidigh & P. J. S. Boaden, pp. 215–24. Pergamon, UK.

Estes, J. A., Jameson, R. J. & Rhode, E. B. (1982). Activity and prey selection in the sea otter: influence of population status on community structure. *Am. Nat.* **120**, 242–58.

Fager, E. W. (1968). A sand-bottom epifaunal community of invertebrates in shallow water. *Limnol. Oceanogr.* **13**, 448–64.

Fedra, K. (1977). Structural features of a north Adriatic benthic community. In *Biology of Benthic Organisms*, ed. B. F. Keegan, P. O'Céidigh & P. J. S. Boaden, pp. 233–46. Pergamon, UK.

Field, J. G., Jarman, N. J., Dieckmann, G. S., Griffiths, C. L., Velimirov, B. & Zoutendyk, P. (1977). Sun, waves, seaweed and lobsters: the dynamics of a west coast kelp bed. *S. Afr. J. Sci.* **73**, 7–10.

Field, J. G., Griffiths, C. L., Griffiths, R. J., Jarman, N., Zoutendyk, P., Velimirov, B. & Bowes, A. (1980). Variation in structure and biomass of kelp communities along the south-west cape coast. *Trans. R. Soc. S. Afr.* **44**, 145–203.

Flemming, N. C. & Stride, A. H. (1967). Basal sand and gravel patches with separate indications of tidal current and storm-wave paths, near Plymouth. *J. mar. biol. Ass. UK*, **47**, 433–44.

Fricke, H. W. (1973). Behaviour as part of ecological adaptation – *in situ* studies in a coral reef. *Helgoländer wiss. Meeresunters*. **24**, 120–44.

Fricke, H. W. (1975). The role of behaviour in marine symbiotic animals. *Symp. Soc. exp. Biol.* **29**, 581–93.

Fricke, H. W. & Fricke, S. (1977). Monogamy and sex change by aggressive dominance in coral reef fish. *Nature, Lond.* **266**, 830–2.

Geister, J. (1977). The influence of wave exposure on the ecological zonation of Caribbean coral reefs. In *Proceedings, Third International Coral Reef Symposium*, vol. 1, pp. 23–29. Rosenstiel School of Marine and Atmospheric Studies, Miami.

Gerard, V. A. & Mann, K. H. (1979). Growth and production of *Laminaria longicruris* (Phaeophyta) populations exposed to different intensities of water movement. *J. Phycol.* **15**, 33–41.

Ghelardi, R. J. (1971). Species structure of the holdfast community. In *The Biology of Giant Kelp Beds (Macrocystis) in California*, ed. W. J. North, pp. 381–420. *Beihefte zur Nova Hedwigia*, vol. 32.

Goreau, T. F. & Goreau, N. I. (1973). The ecology of Jamaican coral reefs. II Geomorphology, zonation and sedimentary phases. *Bull mar. Sci.* **23**, 399–464.

Gorlich, D. L., Atkins, P. D. & Losey, G. S. Jr. (1978). Cleaning stations as water holes, garbage dumps and sites for the evolution of reciprocal altruism? *Am. Nat.* **112**, 341–53.

Goldman, B. & Talbot, F. H. (1976). Aspects of the ecology of coral reef fishes. In *Biology and Geology of Coral Reefs*, vol. III, Biology 2, ed. O. A. Jones & R. Endean, pp. 125–54. Academic Press, New York.

Graus, R. R. & Macintyre, I. G. (1976). Light control of growth form in colonial reef corals: computer simulation. *Science (Wash.)*, **193**, 895–7.

Greenway, M. (1976). The grazing of *Thalassia testudinum* in Kingston Harbour, Jamaica. *Aquat. Bot.* **2**, 117–26.

Gulliksen, B. (1980). The macrobenthic rocky-bottom fauna of Borgenfjorden, North-Tröndelag, Norway. *Sarsia*, **65**, 115–38.

Gulliksen, B., Haug, T. & Sandnes, O. K. (1980). Benthic macrofauna on new and old lava grounds at Jan Mayen. *Sarsia*, **65**, 137–48.

Hartnoll, R. G. (1967). An investigation of the movement of the scallop *Pecten maximus*. *Helgoländer wiss. Meeresunters*. **15**, 523–33.

Highsmith, R. C. (1979). Coral growth rates and environmental control of density banding. *J. exp. mar. Biol. Ecol.* **37**, 105–25.

Highsmith, R. C. (1980). Geographic patterns of coral bioerosion: a productivity hypothesis. *J. exp. mar. Biol. Ecol.* **46**, 177–96.

Hiscock, K. & Hoare, R. (1975). The ecology of sublittoral communities at Abereiddy quarry, Pembrokeshire. *J. mar. biol. Ass. UK*, **55**, 833–64.

Hughes, R. G. (1975). The distribution of epizoites on the hydroid *Nemertesia antennina* (L.). *J. mar. biol. Ass. UK*, **55**, 275–94.

Huston, M. (1979). A general hypothesis of species diversity. *Am. Nat.* **113**, 81–101.

Itzkowitz, M. (1974). A behavioural reconnaissance of some Jamaican reef fishes. *Zool. J. Linn. Soc.* **55**, 87–118.

Itzkowitz, M. (1977). Spatial organization of the Jamaican damselfish community. *J. exp. mar. Biol. Ecol.* **28**, 217–41.

Jackson, J. C. B. (1977). Competition on marine hard substrata: the adaptive significance of solitary and colonial strategies. *Am. Nat.* **111**, 743–67.

Jackson, J. C. B. (1979). Morphological strategies of sessile animals. In *Biology and Systematics of Colonial Organisms*, Systematics Association Special Volume No. 11, ed. G. Larwood & B. R. Rosen, pp. 499–555. Academic Press, London & New York.

Johannes, R. E. (1974). Sources of nutritional energy for reef corals. In *Proceedings of the Second International Coral Reef Symposium*, vol. 1, pp. 133–8. Great Barrier Reef Committee, Brisbane.

Johannes, R. E. & Tepley, L. (1974). Examination of feeding of the reef coral *Porites lobata in situ* using time lapse photography. In *Proceedings of the Second International Coral Reef Symposium*, vol. 1, pp. 127–32. Great Barrier Reef Committee, Brisbane.

Johnson, J. K. (1972). Effect of turbidity on the rate of filtration and growth of the slipper limpet, *Crepidula fornicata* Lamark, 1799. *Veliger*, **14**, 315–20.

Jones, D. J. (1971). Ecological studies on macroinvertebrate populations associated with polluted kelp forests in the North Sea. *Helgoländer wiss. Meeresunters.* **22**, 417–41.

Jones, O. A. & Endean, R. (eds.) (1973a, b, 1976, 1977). *Biology and Geology of coral Reefs*. I, Geology 1; II, Biology 1; III, Biology 2; IV, Geology 2. Academic Press, London, New York.

Kain, J. M. (1963). Aspects of the biology of *Laminaria hyperborea* II. Age, weight and length. *J. mar. biol. Ass. UK*, **43**, 129–51.

Kain, J. M. (1977). The biology of *Laminaria hyperborea* X. The effect of depth on some populations. *J. mar. biol. Ass. UK*, **57**, 587–608.

Kain, J. M. (1979). A view of the genus *Laminaria*. *Oceanogr. mar. Biol.* **17**, 101–61.

Karlson, R. (1978). Predation and space utilization patterns in a marine epifaunal community. *J. exp. mar. Biol. Ecol.* **31**, 225–39.

Kaufman, L. (1977). The three spot damselfish: effects on benthic biota of Caribbean coral reefs. In *Proceedings, Third International Coral Reef Symposium*, vol. 1, pp. 559–64. Rosenstiel School of Marine and Atmospheric Studies, Miami.

Koehl, M. A. R. (1976). Mechanical design in sea anemones. In *Coelenterate Ecology and Behaviour*, ed. G. O. Mackie, pp. 23–32. Plenum Press, New York.

Koehl, M. A. R. (1977a). Effects of sea anemones on the flow forces they encounter. *J. exp. Biol.* **69**, 87–105.

Koehl, M. A. R. (1977b). Mechanical diversity of connective tissue of the body wall of sea anemones. *J. exp. Biol.* **69**, 107–25.

Koehl, M. A. R. (1977c). Mechanical organisation of cantilever-like sessile organisms: sea anemones. *J. exp. Biol.* **69**, 127–42.

Koehl, M. A. R. (1982). Mechanical design of spicule-reinforced connective tissue: stiffness. *J. exp. Biol.* **98**, 239–67.

Koehl, M. A. R. & Wainwright, S. A. (1977). Mechanical adaptations of a giant kelp. *Limnol. Oceanogr.* **22**, 1067–71.

Krumbein, W. E. & Pers, J. N. C. (1974). Diving investigations on biodeterioration by sea urchins in the rocky sublittoral of Helgoland. *Helgoländer wiss. Meeresunters.* **26**, 1–17.

Lang, J. (1973). Interspecific aggression by scleractinian corals. 2. Why the race is not only to the swift. *Bull. mar. Sci.* **23**, 260–79.

Lasker, H. R. (1981). A comparison of the particulate feeding abilities of three species of gorgonian soft coral. *Mar. Ecol. Prog. Ser.* **5**, 61–7.

Lawrence, J. M. (1975). On the relationships between marine plants and sea urchins. *Oceanogr. mar. Biol.* **13**, 213–86.

Leighton, D. L. (1971). Grazing activities of benthic invertebrates in kelp beds. In *The Biology of Giant Kelp Beds (Macrocystis) in California*, ed. W. J. North, pp. 421–53. *Beihefte zur Nova Hedwigia*, vol. 32.

Leversee, G. J. (1976). Flow and feeding in fan-shaped colonies of the gorgonian coral, *Leptogorgia. Biol. Bull. mar. biol. Lab. Woods Hole*, **151**, 344–56.

Lewis, J. B. (1976). Experimental tests of suspension feeding in Atlantic reef corals. *Mar. Biol.* **36**, 147–50.

Lewis, J. B. (1977). Processes of organic production on coral reefs. *Biol. Rev.* **52**, 305–47.

Lewis, J. B. & Price, W. S. (1975). Feeding mechanisms and feeding strategies of Atlantic reef corals. *J. Zool. Lond.* **176**, 527–44.

Limbaugh, C. (1961). Cleaning symbiosis. *Scient. Am.* **205**, 42–9.

Luckhurst, B. E. & Luckhurst, K. (1978). Diurnal space utilization in coral reef fish communities. *Mar. Biol.* **49**, 325–32.

Mann, K. H. (1973). Seaweeds: their productivity and strategy for growth. *Science*, **182**, 975–81.

Mann, K. H. (1977). Destruction of kelp beds by sea urchins: a cyclical phenomenon or irreversible degradation? *Helgoländer wiss. Meeresunters.* **30**, 455–67.

Mann, K. H. (1982). *Ecology of Coastal Waters: A Systems Approach*. Blackwell, Oxford.

Mann, K. H., Jarman, N. & Dieckmann, G. (1979). Development of a method for measuring the productivity of the kelp *Ecklonia maxima* (Osbeck). *Trans. R. Soc. S. Afr.* **44**, 27–41.

Mariscal, R. N. (1972). Behaviour of symbiotic fishes and sea anemones. In *Behaviour of Marine Animals*, vol. 2, *Vertebrates*, ed. H. E. Winn & B. L. Olla, pp. 327–60. Plenum Press, New York & London.

Meyer, D. L. (1982). Food and feeding mechanisms: Crinozoa. In *Echinoderm Nutrition*, ed. M. Jangoux & J. M. Lawrence, pp. 25–42. Balkema, Rotterdam.

Miller, R. J. & Colodey, A. G. (1983). Widespread mass mortalities of the green sea urchin in Nova Scotia, Canada. *Mar. Biol.* **73**, 263–7.

Miller, R. J., Mann, K. H. & Scarratt, D. J. (1971). Production potential of a seaweed–lobster community in eastern Canada. *J. Fish. Res. Bd. Can.* **28**, 1733–8.

Mills, E. L. (1969). The community concept in marine zoology, with comments on continua and instability in some marine communities: a review. *J. Fish. Res. Bd. Can.* **26**, 1415–28.

Moore, P. G. (1977). Inorganic particulate suspensions in the sea and their effects on marine animals. *Oceanogr. mar. Biol.* **15**, 225–363.

Muzik, K. & Wainwright, S. A. (1977). Morphology and habitat of five Fijian sea fans. *Bull. mar. Sci.* **27**, 308–37.

Neudecker, S. (1977). Transplant experiments to test the effect of fish grazing on coral distribution. In *Proceedings, Third International Coral Reef Symposium*, vol. 1, pp. 317–23. Rosenstiel School of Marine and Atmospheric Studies, Miami.

North, W. J. (ed.) (1971). The biology of giant kelp beds (*Macrocystis*) in California. *Beihefte zur Nova Hedwigia*, **32**, 1–600.

Ogden, J. C. & Buckman, N. S. (1973). Movement, foraging groups, and diurnal migrations of the striped parrot fish *Scarus croicensis* Bloch (Scaridae). *Ecology*, **54**, 589–96.

Ogden, J. C. & Ehrlich, P. R. (1977). The behaviour of heterotypic resting schools of juvenile grunts (Pomadasyidae). *Mar. Biol.* **42**, 273–80.

Ormond, R. F. G. (1980). Aggressive mimicry and other interspecific feeding associations among Red Sea coral reef predators. *J. Zool. Lond.* **191**, 247–62.

Patton, W. K. (1974). Community structure among the animals inhabiting the coral *Pocillopora damicornis* at Heron Island, Australia. In *Symbiosis in the Sea*, ed. W. B. Vernberg, pp. 219–43. University of South Carolina Press.

Patton, W. K. (1976). Animal associates of living reef corals. In *Biology and Geology of Coral Reefs*, vol. III, *Biology 2*, ed. O. A. Jones & R. Endean, pp. 1–36. Academic Press, London & New York.

Pearson, R. G. (1981). Recovery and recolonization of coral reefs. *Mar. Ecol. Prog. Ser.* **4**, 105–22.

Pequegnat, W. E. (1964). The epifauna of a California siltstone reef. *Ecology*, **45**, 272–83.

Porter, J. W. (1974). Zooplankton feeding by the Caribbean reef-building coral *Montastrea cavernosa*. In *Proceedings of the Second International Coral Reef Symposium*, vol. 1, pp. 111–26. Great Barrier Reef Committee, Brisbane.

Porter, J. W. (1976). Autotrophy, heterotrophy, and resource partitioning in Caribbean reef-building corals. *Am. Nat.* **110**, 731–42.

Potts, F. A. (1915). *Hapalocarcinus*, the gall forming crab, with some notes on the related genus *Cryptochirus*. *Publs. Carnegie Instn.* **212**, 33–69.

Potts, G. W. (1970). The schooling ethology of *Lutianus monostigma* (Pisces) in the shallow reef environment of Aldabra. *J. Zool. Lond.* **161**, 223–35.

Potts, G. W. (1973). The ethology of *Labroides dimidiatus* (Cuv. & Val.) (Labridae, Pisces) on Aldabra. *Anim. Behav.* **21**, 250–91.

Potts, G. W. (1980). The predatory behaviour of *Caranx melampygus* (Pisces) in the channel environment of Aldabra Atoll (Indian Ocean). *J. Zool. Lond.* **192**, 323–50.

Potts, G. W. (1981). Behavioural interactions between the Carangidae (Pisces) and their prey on the fore-reef slope of Aldabra, with notes on other predators. *J. Zool. Lond.* **195**, 385–404.

Quast, J. C. (1971a). Fish fauna of the rocky inshore zone. In *The Biology of Giant Kelp Beds (*Macrocystis*) in California*, ed. W. J. North, pp. 481–507. *Beihefte zur Nova Hedwigia*, vol. 32.

Quast, J. C. (1971b). Estimates of the populations and standing crop of kelp bed fishes. In *The Biology of Giant Kelp Beds (*Macrocystis*) in California*, ed. W. J. North, pp. 509–40. *Beihefte zur Nova Hedwigia*, vol. 32.

Richardson, C. A., Dustan, P. & Lang, J. C. (1979). Maintenance of living space by sweeper tentacles of *Montastrea cavernosa*, a Caribbean reef coral. *Mar. Biol.* **55**, 181–6.

Robertson, D. R. & Choat, J. H. (1974). Protogynous hermaphroditism and social systems in labrid fish. In *Proceedings of the Second International Coral Reef Symposium*, vol. 1, pp. 217–25. Great Barrier Reef Committee, Brisbane.

Robins, M. W. (1968). The ecology of *Alcyonium* species in the Scilly Isles. *Underwater Assoc. Rep.* **1968**, 67–71.

Rogers, C. S. (1979). The productivity of San Cristobal reef, Puerto Rico. *Limnol. Oceanogr.* **24**, 342–9.

Sale, P. F. (1980). The ecology of fishes on coral reefs, *Oceanogr. mar. Biol.* **18**, 367–421.

Shapiro, D. Y. (1977). The structure and growth of social groups of the hermaphroditic fish *Anthias squamipinnis* (Peters). In *Proceedings, Third International Coral Reef Symposium*, vol. 1, pp. 571–7. Rosenstiel School of Marine and Atmospheric Studies, Miami.

Shapiro, D. Y. (1981). Size, maturation and the social control of sex reversal in the coral reef fish *Anthias squamipinnis*. *J. Zool. Lond.* **193**, 105–28.

Shaw, J. K. & Hopkins, T. S. (1977). The distribution of the family Hapalocarcinidae (Decapoda, Brachyura) on the Florida Middle Ground with a description of *Pseudocryptochirus hypostegus* n. sp. In *Proceedings, Third International Coral Reef Symposium*, vol. 1, pp. 177–83. Rosenstiel School of Marine and Atmospheric Studies, Miami.

Sheppard, C. R. C. (1979). Interspecific aggression between reef corals with reference to their distribution. *Mar. Ecol. Prog. Ser.* **1**, 237–47.

Sheppard, C. R. C. (1981). Illumination of the coral community beneath tabular *Acropora* species. *Mar. Biol.* **64**, 53–8.

Sloan, N. A. (1981). Observations on an aggregation of the starfish *Asterias rubens* L. in Morecambe Bay, Lancashire, England. *J. nat. Hist.* **15**, 407–18.

Stachowitsch, M. (1977). The hermit crab microbiocoenosis – the role of motile secondary hard-bottom elements in a north Adriatic benthic community. In *Biology of Benthic Organisms*, ed. B. F. Keegan, P. O'Céidigh & P. J. S. Boaden, pp. 549–58. Pergamon, UK.

Stearn, C. W. & Scoffin, T. P. (1977). Carbonate budget of a fringing reef, Barbados. In *Proceedings, Third International Coral Reef Symposium*, vol. 2, pp. 471–6. Rosenstiel School of Marine and Atmospheric Studies, Miami.

Stearn, C. W., Scoffin, T. P. & Martindale, W. (1977). Calcium carbonate budget of a fringing reef on the west coast of Barbados. Pt. I Zonation and productivity. *Bull. mar. Sci.* **27**, 479–510.

Thresher, R. E. (1977). Ecological determinants of social organisation of reef fishes. In *Proceedings, Third International Coral Reef Symposium*, vol. 1, pp. 553–7. Rosenstiel School of Marine and Atmospheric Studies, Miami.

Velimirov, B. (1973). Orientation in the sea fan *Eunicella cavolinii* related to water movement. *Helgoländer wiss. Meeresunters.* **24**, 163–73.

Vogel, S. (1974). Current-induced flow through the sponge *Halichondria*. *Biol. Bull. mar. biol. Lab. Woods Hole*, **147**, 443–56.

Vogel, S. (1978). Organisms that capture currents. *Scient. Am.* **239**, 108–17.

Wainwright, S. A. & Dillon, J. R. (1969). On the orientation of sea fans (genus *Gorgonia*). *Biol. Bull. mar. biol. Lab. Woods Hole*, **136**, 130–9.

Wainwright, S. A., Biggs, W. D., Currey, J. D. & Gosline, J. M. (1976). *Mechanical Design in Organisms*. Edward Arnold, London.

Warner, G. F. (1971). On the ecology of a dense bed of the brittle-star *Ophiothrix fragilis*. *J. mar. biol. Ass. UK*, **51**, 267–82.

Warner, G. F. (1977). On the shapes of passive suspension feeders. In *Biology of Benthic Organisms*, ed. B. F. Keegan, P. O'Céidigh & P. J. S. Boaden, pp. 567–76. Pergamon, UK.

Warner, G. F. (1979). Aggregation in echinoderms. In *Biology and Systematics of Colonial Organisms*, Systematics Association Special Volume No. 11, ed. G. Larwood & B. R. Rosen, pp. 375–96. Academic Press, London & New York.

Warner, G. F. (1981). Species descriptions and ecological observations of black corals (Antipatharia) from Trinidad. *Bull. mar. Sci.* **31**, 147–61.

Warner, G. F. (1982). Food and feeding mechanisms: Ophiuroidea. In *Echinoderm Nutrition*, ed. M. Jangoux & J. M. Lawrence, pp. 161–81. Balkema, Rotterdam.

Warner, G. F. & Moore, C. A. M. (1984). Ecological studies in the marine blue-holes of Andros Island, Bahamas. *Cave Science*, **11**, in press.

Warner, G. F. & Woodley, J. D. (1975). Suspension feeding in the brittle-star *Ophiothrix fragilis. J. mar. biol. Ass. UK*, **55**, 199–210.

Warner, R. R. & Downs, I. F. (1977). Comparative life histories: growth *vs.* reproduction in normal males and sex-changing hermaphrodites of the striped parrot fish, *Scarus croicensis.* In *Proceedings, Third International Coral Reef Symposium*, vol. 1, pp. 275–81. Rosenstiel School of Marine and Atmospheric Studies, Miami.

Wellington, G. M. (1982). Depth zonation of corals in the Gulf of Panama: control and facilitation by resident reef fishes. *Ecol. Monogr.* **52**, 223–41.

Wing, B. L. & Clendenning, K. A. (1971). Kelp surfaces and associated invertebrates. In *The Biology of Giant Kelp Beds (*Macrocystis*) in California*, ed. W. J. North, pp. 319–41. *Beihefte zur Nova Hedwigia*, vol. 32.

Young, C. M. & Braithwaite, L. F. (1980). Orientation and current-induced flow in the stalked ascidian *Styella montereyensis. Biol. Bull. mar. biol. Lab. Woods Hole*, **159**, 428–40.

INDEX

Page references in bold type indicate major entries, those in italic indicate illustrations